RATIONALIZING MIGRATION DECISIONS

RATIONALIZING MIGRATION DECISIONS

Dedicated to my parents,
who being direst victims of the liberation war in 1971 by losing all
our assets and properties, kept praying for our enlightened life

Rationalizing Migration Decisions
Labour Migrants in East and South-East Asia

AKM AHSAN ULLAH
The American University in Cairo, Egypt

Routledge
Taylor & Francis Group

LONDON AND NEW YORK

First published 2010 by Ashgate Publishing

2 Park Square, Milton Park, Abingdon, Oxon OX14 4RN
711 Third Avenue, New York, NY 10017, USA

Routledge is an imprint of the Taylor & Francis Group, an informa business

First issued in paperback 2016

British Library Cataloguing in Publication Data
Ahsan Ullah, A. K. M.
  Rationalizing migration decisions : labour migrants in East
  and South-east Asia.
  1. Emigration and immigration--Psychological aspects.
  2. Emigration and immigration--Economic aspects.
  3. Foreign workers, Bangladeshi--China--Hong Kong--
  Social conditions--21st century. 4. Foreign workers,
  Bangladeshi--Malaysia--Social conditions--21st century.
  5. Foreign workers, Bangladeshi--China--Hong Kong--
  Attitudes. 6. Foreign workers, Bangladeshi--Malaysia--
  Attitudes. 7. Bangladesh--Emigration and immigration.
  8. Hong Kong (China)--Emigration and immigration.
  9. Malaysia--Emigration and immigration.
  I. Title
  304.8'512505492-dc22

Library of Congress Cataloging-in-Publication Data
Ahsan Ullah, A. K. M.
  Rationalizing migration decisions : labour migrants in East and South-East Asia / by A.K.M.
Ahsan Ullah.
    p. cm.
  Includes bibliographical references and index.
  ISBN 978-1-4094-0513-9 (hardback)
  1. Migrant labor--Bangladesh. 2. Foreign workers--China--Hong Kong. 3. Foreign
workers--Malaysia. 4. Bangladesh--Emigration and immigration. 5. Hong Kong (China)--
Emigration and immigration. 6. Malaysia--Emigration and immigration. I. Title.
  HD5856.B36A47 2010
  331.6'2095--dc22

                                                                              2010006283

ISBN 978-1-4094-0513-9 (hbk)
ISBN 978-1-138-26074-0 (pbk)

Contents

Contents

List of Figures

Figures

Maps

Photo

List of Tables

Foreword

Though yet in its early days, the Third Millennium looks destined to be, in sheer numbers, the age of the greatest human mobility in recorded history. Today there are nearly 1 billion migrants in the world – one in every six or seven people – made up of more than 200 million international migrants and 740 million internal migrants. Growing labour demands and deficits in the industrialized world due to ageing populations and dramatically declining birth rates, combined with rapid population growth, slow job creation and surplus labour in the developing world, will insure the inevitability of human migration throughout this century.

Rationalizing Migration Decisions: Labour Migrants in East and South-East Asia examines the decisions people make to leave their homes and migrate abroad. In doing so, the book examines the economic and social costs of migration, versus the benefits of future income along with the inherent risks and adversities faced along the way. Among the points raised in the study is the reality that while the decision to migrate may well be an individual's choice, the migration process extends well beyond the control of anyone individual, including as it does, migration policies in origin and destination countries, economic factors, and the support available in the new host community.

The rich ethnographic content of the book seeks to answer the question of why migrants leave their countries and remain in destination countries despite the risk of considerable social, psychological and economic costs, including vulnerability as undocumented workers, and incurring financial risks that often leave them and their families heavily indebted.

The author attempts to answer this question through examining the migration process in its pre-departure and post-arrival phases. The outcome of this investigation is a characterization of the flexible nature of individuals as they go about their migration choices. Contrary to characterizations that migrants are unaware of the conditions and consequences of their decisions, the author argues that individual migrants arm themselves with information and knowledge of the difficult journey and its losses and gains and trade-offs, even if this means working in precarious circumstances.

The author's use of quantitative and qualitative methodologies straddles sociological and economic schools of thought on migration, and in doing so endears this work to a wider audience and contributes to the on-going global migration dialogue.

William Lacy Swing, Director General,
International Organization for Migration (IOM), Geneva, Switzerland

Preface

One of the most important components of contemporary globalization is the influx and movement of people around the world. Indeed, the growing significance of migration has engendered an abundance of studies exploring the many facets and dynamics of population migration. With the increase in volume of migration flows worldwide, Bangladesh has likewise seen considerable expansion in its flows of migration. Today, migration has become a public policy matter for Bangladesh and for the countries receiving Bangladeshi migrants – among which are Hong Kong and Malaysia which are addressed in this book.

A significant group of Bangladeshi migrants have been in Hong Kong for the last three decades, despite the fact that this metropolis is not considered to be a destination of first choice. Malaysia, on the other hand, has been a major host to Bangladeshi migrants, under the auspices of the labour export agreements of the Nineties, hiring around 50,000 skilled and unskilled workers annually to cover the severe shortage in its labour market.

This book examines how Bangladeshi migrant workers in Hong Kong (BRHKs) and Malaysia (BRMs) go about their decisions to migrate and how they rationalize their decision post-migration. In order to do this, the research investigates two perspectives in the migration process: the first is pre-migration decision-making; and the second is rationalizing post-migration experiences. While decisions about working overseas are often based on expectations and promises of increased opportunities, economic gains and, eventually, a better future, such assumptions may not always be realized. Even when they are, mounting evidence indicates that migrants also suffer numerous adversities tied to the migration process. Herein lies the key argument of the book: Despite the countless hardships that the migrants themselves go through and continue to experience post-migration, they nevertheless justify their migration decisions positively. While these justifications may be seen from an academic standpoint as subjective and may go beyond the bounds of quantitative calculations, the fact remains that the migrants themselves desire to shape their own narratives. This book records these narratives – many dealing with myriad hardships ranging from emotional to financial and physical – taken from migrants' first-hand experiences.

Apart from the financial costs incurred by the migration process, social and psychological costs are also paid, often at a steep price, although these are difficult to measure in absolute terms. This situation places migrant workers in a state where, in order to protect their own sense of well-being, they justify their migration decision. While theories and the treatment of migration issues in various research has addressed the rationalization of migration decision previously, most of this work has been presented in a dispersed manner. This book is an attempt to come up with a unified understanding of the process of rationalization of the migration decision.

While different circumstances have different effects on an individual's migration choice, economic considerations have always registered as highly influential throughout the migration process. In addition, networking, financial cost, living and working conditions, income benefits, remittances and their impact on well-being have been taken into account as other key variables of rationalization.

This book proposes an expansion of the network theory by debunking conventionally-accepted wisdom, using strong empirical data. The book does not cease to talk about the exorbitant costs that migration incurs; rather it sheds light on the process' long term implications for migrants, who often have to borrow in order to finance their migration. Comparative data between Malaysia and Hong Kong is analyzed against a number of variables. The research indicates that the migrants in Malaysia take up riskier and more dangerous jobs with less income benefits than do the migrants in Hong Kong. While both the migrants in Hong Kong and those in Malaysia were subjected to varying degrees and forms of vulnerability and exploitation, the level of suffering experienced among the migrants in Malaysia was higher than that experienced by those in Hong Kong. In addition, despite being offered similar categories of work, the income level of the migrants in Hong Kong tended to be significantly higher than that of those in Malaysia. Similarly, the Hong Kong migrant workers remitted higher amounts than the migrants in Malaysia. This study has demonstrated that a large proportion of the remittances being transferred back to Bangladesh is done so through informal channels due to the relative inefficiency and complicated procedures associated with formal systems. Although these remittances significantly contribute to the Bangladeshi GDP at the macro-level, they present a different story at the micro-level: The study reveals that a major portion of migrants' remittance is usually spent on unproductive expenses.

All in all, the book details the common sentiment of migrants: that despite the adversities that they suffer, they still intend to stay abroad. This decision has logic to it: while they may only reach a break-even point, economically, while staying in their destination countries, compounded by being at-risk legally, physically and emotionally, this remains a better choice than returning home empty-handed with no foreseeable way of earning money to recoup their expenses or buy back the land that they have pawned or sold. This need to recover the investment they made in order to finance their migration accounts for the length of time that these migrants remain in their host countries, and for their subconscious motivation to rationalize their initial decision to migrate.

AKM Ahsan Ullah
2010

Acknowledgements

In any research, the respondents' contribution is the most significant part. My thanks and gratitude go to the respondents who provided information for this book. Their help in providing information has been simply immeasurable without which the research would not have been possible to accomplish. My thanks to them because of their spontaneity in making a lot of insights available to me I did not ever expect.

In order to keep me focused on my research Dr Vivienne Wee never forgot to catch me even on her way home. She was concerned not only with academic guidance but also that I was at peace to work and to stay in Hong Kong during my research. Her immense support in various aspects is simply incalculable. I would like to express my greatest appreciation to her for the encouragement, and insightful and scholarly guidance she extended incessantly during the whole period of my research.

I would like to express my greatest appreciation to my scholarly teacher Professor Kevin Hewison, whose influence raised in me the interest in the growing issues of migration scholarship. While writing this acknowledgement, I am reminded of the steps of proposal development from issues of the NGOs and their role in development, on which I was accepted to do my research, to my shift of focus to migration. I am reminded, how I made up my mind so easily to work on the issue of migration, to me a fresh discipline indeed. Professor Hewison has been a guide behind all these. When I got him as my supervisor, he introduced me to the vast amount of theories related to migration studies.

I am indebted to Dr Ahmed Shafiqul Huque, McMaster University, who has done more than I expected. I am just unable to explain in words how he continued to give me encouragements on my work. His contribution is gratefully acknowledged. Comments made by him on the manuscript had pushed me to jump higher and higher.

I render my thanks to the City University of Hong Kong for continuing my scholarship, for offering me research grants, a conducive environment for doing research, conference grants and research tuition scholarships. I remember all the facilities given to me throughout my research period. My thanks go to the Department of Asian and International Studies (AIS) and the Southeast Asia Research Centre (SEARC) for the immense support extended toward my research. The City University of Hong Kong shall remain in my heart until my last breath.

I have learned a great deal from several friends who have shared with me their insights, experiences, criticisms, and comments. Both directly and indirectly, their input is no doubt reflected in this book. I am grateful to all of them. It would simply be an injustice if a few names such as Dr Mizanur Rahman, NUS, Singapore; Dr Mallik Akram Hossain, Department of Geography, Rajshahi University, Professor Taiabur Rahman, Dhaka University who extended support in many ways are not mentioned. I am grateful to Liyam Eloul for her excellent editorial support.

I thank the Centre for Development Research (ZEF), University of Bonn, Germany for offering me the scholarship for my pre-doctoral studies. Anabelle Ragsag is specially acknowledged for her immense support in many respects, especially by giving valuable comments. Her contribution to my work is ever acknowledged. I am grateful to the Canadian Institutes of Health Research (CIHR) for offering me a Post-Doctoral Fellowship for three years.

At this stage, I must remember a few names who have always been supportive in pursuing my higher studies; they are Professor M. Asaduzzaman (Late), ex-Chairman of the University Grants Commission, Bangladesh, Professor Salahuddin Aminuzzaman, University of Dhaka; Professor Jayant K. Routray, Asian Institute of Technology (AIT), Thailand; Dr Sanzidur Rahman, Plymouth University, UK and Dr Katherine Gould-Martin, Bard College, USA and Professor Ronald Labonté, Canada Research Chair, Globalization/Health Equity, University of Ottawa, Canada.

My family members whose prayers and patience have made it possible to finish my research are gratefully acknowledged. Many of my family members made significant sacrifices in the course of my long academic journey. I record my boundless thanks for the patience shown to me, and seek apologies for the pain I caused to them.

List of Abbreviations

ACTFORM	Action Network for Migrants
ADB	Annual Development Budget
ADB	Asian Development Bank
ADWU	Asian Domestic Workers Union
ALICO	American life insurance company
AMC	Asia Migration Centre
APMRN	Asia Pacific Migration Research Network
ARCM	Asian Research Centre for Migration
ASEAN	Association of Southeast Asian Nations
ATM	Automatic teller machine
BBS	Bangladesh Bureau of Statistics
BFM	Bangladeshi female migrants
BMET	Bureau of Manpower Employment and Training
BoP	Balance of payment
CAPSTRAN	Centre for Asia Pacific Social Transformation Studies
CATW	Trafficking in Women and Prostitution in the Asia Pacific
CBA	Cost-benefit analysis
CIA	Central Intelligence Agency
CIMW	Center for Indonesian Migrant Workers
CMA	Center for Migrants Advocacy
CMD	Centre for Migration and development
CMR	Coalition for Migrants Rights
DR	Dependency ratio
EAP	Economically active population
EPZ	Export processing zone
FAO	Food and agriculture organization of the United Nations
FATF	Financial action task force on money laundering
FDH	Foreign domestic helper
FGS	Focus on the Global South
FWCC	Foreign workers cabinet committee
GA	Gravity approach
GCIM	Global Commission on International Migration
GDI	Gross development index
GDP	Gross development product
GI	Gini Index
GNP	Gross National Product
GoB	Government of Bangladesh

HC	Human capital
HCR	Head count ratio
HDI	Human Development Index
HDR	Human Development Report
HKD	Hong Kong Dollar
HKID	Hong Kong Identity card
HKRs	Hong Kong Respondents
HTM	Harris Todaro Model
ILO	International Labour Organization
IMF	International Monetary Fund
INSTRAW	International Research and Training Institute for the Advancement of Women
IOM	International Organization for Migration
JCMK	Joint Committee for Migrant Workers in Korea
JS – APMDD	Jubilee South – Asia Pacific Movement on Debt and Development
KAKAMMPI	Kapisananng Kamag-anakan ng mga Manggagawang Migranteng Pilipino
KL	Kuala Lumpur
KSA	Kingdom of Saudi Arabia
LDC	Less developed country
MDG	Millennium Development Goal
MEC	Middle East Countries
MFA	Migrant Forum in Asia
MFI	Migrant Forum India
MMN	Mekong Migration Network
MNC	Multi-national corporation
MRs	Malaysia Respondents
MTO	Money transfer operator
MTR	Mass rapid transport
NEP	New economic policy
NGO	Non governmental organization
NIC	Newly industrialized countries
OECD	Organization for Economic Co-operation and Development
PEST	Political, economic, social and technological
PPF	Production possibility frontier
PRIO	International Peace Research Institute, Oslo
RM	Malaysian Ringgit
RMMRU	Refugee and Migratory Movements Research Unit
RUD	Rural urban disparity
RUM	Rural urban migration
SAMReN	South Asian Migration Research Network
SEAC	Southeast Asian Countries
SEARC	Southeast Asia Research Centre

SLIS	Special labour importation scheme
SLS	Supplementary labour scheme
SMC	Scalabrini Migration centre
SPG	Squared poverty gap
SPSS	Statistical package for social sciences
SWOT	Strength, Weakness, Opportunity and Threat
TI	Transparency International
TST	Tsim Sha Tsui
UBINIG	Unnayan Bikalpo Niti Nirdharoni Gobeshona
UNDP	United Nations Development Programme
UNESCAP	United Nations Economic and Social Commission for Asia and the Pacific
UNESCO	United Nations Educational, Scientific and Cultural Organization
UNFPA	United Nations Population Fund
USAID	United States Agency for International Development
WEDPRO	Women's Education, Development, Productivity and Research Organization
WES	Wage earner's scheme
WMI	Weighted Mean Index
WST	World system theory

Equivalence

1.00 USD (United States Dollars)	70.7000 BDT (Bangladesh Taka)
1.00 USD	7.77092 HKD (Hong Kong Dollars)
1.00 USD	3.55610 MYR (Malaysia Ringgits)
1.00 HKD (Hong Kong Dollars)	0.457718 MYR
1.00 HKD	9.09759 BDT
1.00 MYR (Malaysia Ringgits)	19.8774 BDT

Source: Universal currency converter, 5 December 2009.

PART I
Setting the Scene

Chapter 1

Introduction: Migration as a Twenty-First Century Issue

Globalization, an academic buzzword of the early twenty-first century, is a process that many of its proponents trumpet as providing greater mobility of people, capital and information, and as being comprehensively beneficial (Kahanec and Zimmermann 2008, Beaverstock and Boardwell 2000, Lewellen 2002). This approach has not gone unchallenged, and there have been a variety of globalization discourses which have created a massive amount of literature, both academic and popular (Yang 2003). With recent technological developments – especially in transport and digital technologies – the world has seen unprecedented advances in connectivity. Some even argue that these developments presage an era where the people of the world will be incorporated into a single global society (Guy 2004, Martin 2001, Cwerner 2001, Hugo 1998). Some analysts argue that this process has scaled up dramatically in the last two decades in terms of the ways that people are able to travel, communicate, and conduct business internationally (Haque 2004, Friedrichs 2002, Harcourt 2003).

With this growing interconnectedness, international migration has become very much a part of the process of globalization (Findlay et al. 1996, Li, Findlay and Jones 1998, Iredale, Hawksley and Castles 2003 and 2003a, Hewison and Young 2006). Even though it is difficult to obtain precise statistics on migration – in particular for the case of migrant labourers – either by region or globally, the estimates that are available suggest that the rate of growth of the world's migrant population more than doubled in the four decades from the 1960s to the 2000s (ILO 2002). It is also clear that much of this growth originated in developing countries (Hewison 2003 and 1999, Demuth 2000). The number of people who have resettled in a country other than their own is estimated at 200 million[1] worldwide representing three percent of the world population (IOM 2008, UNESCAP 2006).

In recent years, migrants from Asian countries have become highly visible all over the world as specific areas of origin across South, Southeast and East Asia have become more closely linked to overseas destinations as part of the wider process of globalization (Appleyard 1989, Skeldon 2000, Lucas 2001, Lowell

---

1 In 2000, of these, about 158 million were deemed international migrants; approximately 16 million were refugees fleeing a well-founded fear of persecution; and 900,000 were asylum seekers (Doyle 2004, Balbo and Giovanna 2005, Hefti 1997, Fei 2002, Ghai 2004: 3). According to UNESCAP the estimated total number is 191 million (March–April, 2006: 6).

and Kemper 2004). Labour migration in the Asian region has been historically significant and it continued to proliferate in the late twentieth and early twenty-first centuries (Young 2004, Dannecker 2003). The major countries of origin for migrant workers in Southeast Asia are Burma, Indonesia, Laos, Malaysia, Thailand, the Philippines, Vietnam and Bangladesh, while the receiving countries are Brunei, Malaysia, Singapore, South Korea, Taiwan and Hong Kong (Wee and Sim 2003). A number of these countries may serve as labour exporters, importers or, as in the cases of Thailand and Malaysia, both. Malaysia, for example, has over two million alien workers within its borders and, at the same time, over 200,000 Malaysians working in Japan, Taiwan, Hong Kong, and Singapore (Ministry of Human Resources 2009, Kassim 2001a).

In view of the growing significance of migration for contemporary globalization, numerous studies have been undertaken to illustrate and describe the dynamics of current population mobility. With the escalation of the volume of migration flows worldwide, as will be demonstrated below, labour migration from Bangladesh has expanded significantly and has become a major public policy issue for Bangladesh and for the countries that receive Bangladeshi migrants. Despite this, the contemporary academic literature still evidences relatively little research attention to this issue. This book intends to redress this paucity of research surrounding Bangladeshi labour migration. The book examines how Bangladeshi migrant workers in Hong Kong (HK) and Malaysia rationalize their migration and how they make their decisions to migrate by investigating two perspectives in the migration process. The first is the pre-migration decision-making period; and the second is the post-migration rationalization of the decision to migrate.

This approach is adopted because, while decisions to move overseas for work are often based on expectations and promises of better jobs, opportunities, economic gains and, eventually, a better future, such assumptions may not always be realized, and even when they are, migrants often experience hardship during the pre- and post-migration periods. While examples of migrant workers who gain economic benefits by working overseas are not rare, there is also evidence that their expectations and promises often remain unrealized. These workers face many adversities, such as exorbitant migration costs funded through borrowing against or selling assets, under-payment, threats, illegality or confinement in work places, subhuman living conditions, belated or irregular salaries, and the confiscation of travel documents by employers. This host of factors place migrant workers in a complex situation in which they continuously, consciously or unconsciously, rationalize their migration decision. Theories and research about labour migration have been insufficiently attentive to the importance of rationalization in migration and migrant well-being. Therefore, this book seeks to address these complex issues by using an empirical lens to shed light on this aspect of the migration experience.

International migration and Asia

The proliferation of international labour migration in the Asia Pacific region from the mid-1980s to the onset of the Asian financial crisis in 1997 has influenced the directions, flow and volume of migration from the Philippines, India, Pakistan, Bangladesh, Indonesia and Thailand (Dannecker 2003). Therefore, migration for overseas employment has become a major issue in the economies of a number of Asian countries (IOM 2010, Wickramasekera 1996). During the early period of labour migration, workers migrated to the oil-rich Middle Eastern countries under labour contracts of between one and three years (Huguet 1989, Gunatilleke 1986). An estimate in 2000 confirmed that there were approximately three million Asian workers employed in the oil-rich states. Migrant workers from Bangladesh to the Middle East were concentrated mainly in the Kingdom of Saudi Arabia (KSA), and Kuwait (IOM 2009, 2004, Abdul-Aziz 2001).

With the passage of time, the attraction of Middle Eastern countries faded, largely following a decline in salary base but also due to a sense of disillusionment, some of it related to misbehaviour and brutality on the part of many employers. There were a number of reports of foreign workers being systematically abused and exploited, many living in conditions akin to slavery, as well as female workers becoming victims of sexual abuse and forced confinement (*Daily Star* 2004).[2] Compounding these adversities, the subsequent Gulf Wars compelled potential migrant workers to shift their attention toward industrialized countries in East and Southeast Asia. Coincidentally, countries in Southeast Asia had opened up their borders as their economies grew, and there was an increased demand for workers, just as the growth slowed in the Gulf States (Abdul-Aziz 2001, Kanapathy 2001). Eventually, the Middle Eastern countries were no longer among the main destinations for migrants from Asia, and intraregional migration flows to newly industrialized countries in East Asia replaced the vast outflows to the Middle East that had characterized 1980s (Chantavanich 1997, Castles 2003).

With rapid economic growth in the Asian 'tiger economies' of Singapore, South Korea, Taiwan, Hong Kong and Malaysia, labour shortages grew severe, and the local labour markets failed to adequately meet the demand. The result was often raising wages, which was less sustainable for the newly burgeoning economies. Stouffer (1940) and several others developing macro-level migration equations assume that large regional disparities encourage migration and that regional equalization decreases the migration trend (Green et al. 1995, Ainsaar 2004). Indeed, in Asia, countries in which there was a surplus of labour saw an opportunity to respond to this increased demand (Dannecker 2003). The high unemployment rate in Bangladesh meant that potential migrants were not immune to this new job market. It is important not to discount the critical role of the

---

2   Annually, between 200 and 300 deaths of Bangladeshi workers due to occupational accidents, murder, suicide, drowning, execution and stampede are reported in Saudi Arabia (see also Cheung and Mok 1998, Khan 7 August 2003, *Arab News*).

state in migration decisions: historically, migrants have not responded simply as individuals, states' responses, both official and unofficial, are vital in directing migration flows. In sending countries like Bangladesh, China, the Philippines, Thailand, and Sri Lanka, the state has been active in developing policies for the export of surplus labour in order to earn foreign currency, and many have taken up initiatives to train potential migrant workers, albeit usually on a small scale (Iredale, Hawksley and Castles 2003).

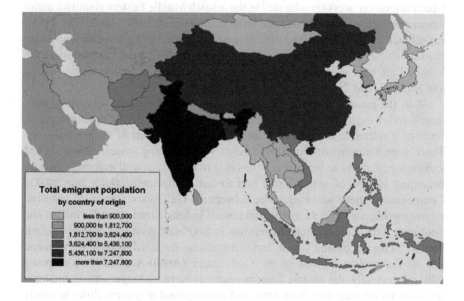

Map 1.1 Map of Asia showing the flow of emigration
Source: IOM 2009.

In order to contextualize my discussion of Bangladeshi migrants in Hong Kong and Malaysia, it is necessary to note broader movements within the Southeast Asian region. The flows of migrants throughout Asia have become complex due to a number of causal factors, including different levels of organized crime involvement. In Malaysia, for example, 'illegals'[3] come from over twenty countries: the majority (85.5 percent) come from Indonesia, due to its geographic proximity and close socio-cultural ties; the second-largest contributor is Myanmar

---

3   Kassim (2001) refers to 'illegal immigrant' in the Malaysian context as an alien entering the country illegally without proper documents or entering legally but overstaying his or her visa, or as a contract worker who defaults on his or her contract, visitors who abuse their visa conditions, or a holder of false travel documents. Even so, illegal migration is often confused with trafficking (IOM 2004).

(4.7 percent), followed by Thailand (3.2 percent), India (2.3 percent), Bangladesh (1.6 percent) and Pakistan (1.2 percent) (Kassim 2001a).

The flow of migration into South East Asia is more intense than that into East Asia. Thailand is a major regional hub for illegal migration. It is a source, destination, transit and facilitation centre all at once (Chantavanich, Germershausen and Beesey 2000), receiving illegal and trafficked people from all of its neighbouring countries. Hong Kong is both a destination country and a site of transit for those trying to sneak into Europe. It is also a recipient of both illegal and trafficked persons brought in for commercial sex, usually from, *inter alia*, China, Thailand, Central Asia, Russia, Vietnam, and the Philippines (Chan and Abdullah 1999). Malaysia is a destination country for trafficked men and women from Indonesia (Chitose 2001), Thailand, Taiwan, South Asia, the Philippines, Cambodia, Myanmar, and China. Singapore is a destination country for young men and women from Malaysia, China, Thailand, South Asia, and the Philippines.

While it has an abundance of both skilled and unskilled labour, Bangladesh's participation in the international labour market has not been as prominent as some other Asian countries, such as the Philippines, Sri Lanka, India, Indonesia and Thailand (Lohrmann 1989, BMET 2009 and 2004). However, there are several major destinations for Bangladeshi migrants in the Asian region, with the highest number in Malaysia. It should be noted that, although known that there are relatively high numbers of migrants in India, in this case it is not necessarily migration for employment: A large number of Bangladeshis leave the country informally for neighbouring India, which serves as a gateway for the large number of mostly women and children who are trafficked to other countries, particularly West Asia (Siddiqui 2004, Ramachandran 2002 and 2005, Gillan 2002). In addition, studies such as Rahman (2003) and Siddiqui (2004) assert that migration from Bangladesh to India is primarily for the purposes of tourism and business.

Bangladesh, with an economy that remains heavily reliant on agriculture, has experienced severe political problems since its inception in 1971 (Zafarullah 1998, GoB, 2009). Independence itself came after a nine-month-long war that is said to have cost some 1,247,000 lives (Siddiqui 2000). Soon after independence, a large number of people from religious minority groups moved to India and many others fled to other countries seeking political asylum.

Moreover, in 1974, floods destroyed much of the country's food crops, causing widespread famine and the deaths of hundreds of thousands of people. Natural disasters have continued to hit Bangladesh with alarming regularity, with 181 recorded, claiming 216,000 lives between 1971 and 2000 (Asian Development Bank 2004, *Daily Prothom Alo* 2004). Despite this, Bangladesh has one of the highest population growth rates in the world with, according to the 2001 census, a total population size of 129.2 million (GoB 2009 and 2004).

In Bangladesh, poverty emerges as the key factor in explaining push forces in migration trends. In the Human Poverty Index, Bangladesh ranked 67th among 78 developing nations. The country's trend in the Human Development Index ranking is equally depressing: Bangladesh stood at 123rd among 146 countries in

1997; 139th among 177 countries in 2003, and 138th among 177 countries in 2004 (UNDP 2004, *Daily Ittefaq* 2004, *Daily Prothom Alo* 2004). Poverty continues to be pervasive (Siddiqui 2000) and all major economic indicators exhibited very slow growth between 2001 and 2008. The profile of the Bangladeshi population indicates a high proportion of young people, with about 44 percent under the age of 15, reaching 50 million in the year 2000. In the same period, the number of fertile women increased to 25 million from the previously reported level of 20 million (Alam and Rahman 2004). Several sources suggest that at least 50 million young people will be looking for employment by the year 2015[4] (*New Nation* 2005).

Bangladesh's high population growth and poverty level are solid indicators of the possibility for increased emigration. However, there are arguments that not all countries with high population growth have correspondingly high emigration rates (Sassen 1988). While poverty is usually seen to be the basic migration push factor, Sassen (1988) points out that not all countries with extensive poverty are emigration countries. Clearly, one factor cannot explain the entire picture. Poverty may create pressure but it does not, in itself, facilitate migration. Therefore, poverty alone does not create the condition of migration. However, the increased likelihood of migration pressures under conditions of poverty, unemployment, and over-population can in no way be denied (Skeldon 2002).

As a general rule, migration takes place from labour surplus and low wage countries or areas to labour deficit and high wage countries or areas (Balbo and Giovanna 2005). Stahl (1991) refined this as a rule for the migration of an unskilled or semi-skilled labour force from poor and labour-abundant countries (or poor and labour-abundant parts of relatively well-developed countries) to relatively rapidly-growing capital-rich countries with a mobile and skilled labour force related to direct foreign investment. Under this rule, Bangladesh figures as a source country for both skilled and unskilled labour. Under the circumstances discussed above, the Bangladeshi people began to migrate. Hereunder compiled from BMET 2009 indicates the trends and factors notable in the main phases of migration. As noted above, there was a mass movement that took place during and immediately after the liberation war of 1971. Indeed, some analysts suggest that Pakistani military operations resulted in 8–10 million refugees crossing the border into India in one of the largest mass movements of people in modern times. After independence, most of these refugees returned, as did approximately 100,000 stranded Bangladeshis from former West Pakistan (LoC 2004). A second phase was formed by a combination of asylum-seeking and labour migration. The two subsequent phases are characterized primarily as labour exports.

---

4   With high unemployment rates (47 percent), the Bangladesh government considers work force export to be a good investment, particularly as currently the country's economy is heavily dependent on migrant's remittances (GoB 2006).

Phases	Major characteristics	Major trend	Major factors
Phase 1 1971–75	Asylum	Worldwide, but India was the primary receiving country	War, poverty, political persecution, pull forces
Phase 2 1975–85	Labour and asylum	Middle East and India	Political persecution, pull forces, and factors of economic migration
Phase 3 1985–95	Labour export	Southeast Asia	Demand, surplus labour, economic growth in neighbouring countries
Phase 4 1995–05	Labour export	Middle East and Southeast Asia	Demand, surplus labour, economic growth in neighbouring countries

As discussed earlier, the mass flow of migrant workers to Middle Eastern countries has declined, however these countries still remain a major labour recruiter for Bangladesh across work categories. A relatively smaller but still significant number are employed in South East Asian countries, in particular Malaysia (INSTRAW/IOM 2000). It should be noted though, that labour exports to Malaysia were dramatically reduced after 2000[5] (BMET 2005), but since mid-2006, the labour agreement with Malaysia came into force, which has renewed the flow of migrant workers from Bangladesh.

Table 1.1 Migration outflow from Bangladesh to selected countries, 2000–2008

Year	Countries					
	Kuwait	Saudi Arabia	Hong Kong	Malaysia	Singapore	South Korea
2000	594	144618	–	17237	11095	990
2001	5341	137248	–	4921	9615	1561
2002	15769	163269	–	85	6856	28
2003	26722	162131	–	28	5304	3771
2004	41108	139031	–	224	6948	215
2005	47029	80425	5000*	2911	9651	223
2006	35775	109513	5000*	20469	20139	992
2007	4212	204112	6500*	273201	38324	39
2008	319	132124	–	131762	56581	1521

Note: '–' indicates no data available; * Estimated according to author's survey, 2005–2006 as well as available print media.
Source: BMET 2009.

---

5 Manpower export to Malaysia was expected to resume after a gap of seven years under a MoU signed in July 2004 (BMET 2005).

Hong Kong and Malaysia as research sites

This section explicates the rationale for selecting the two research sites for this study: Hong Kong and Malaysia.[6] These two destinations present quite different patterns of migration profiles for Bangladeshis. This section also describes the labour demand situation and how these two receiving destinations meet labour demand.

Hong Kong

Strategically located in the rapidly developing Asia-Pacific region, Hong Kong is unique in its historical reliance on migrant workers for economic growth (Athukorala and Manning 2000 and 1999, Skeldon 1990, Ho et al. 1991, Hewison 2006). While its native population is ethnically homogeneous, unlike the other high-performing East Asian countries, Hong Kong has generally been receptive to foreigners, partly reflecting its long colonial history as well as migration patterns from China (Athukorala and Manning 1999, Kowng 1990). Modern-day Hong Kong's population is reasonably diverse and international migration has, in recent years, contributed to this increased level of ethnic diversity (Glavaz and Waldorf 1998). The ethnic break-down for Hong Kong indicates that the Chinese ethnic group still dominates, accounting for around 95 percent[7] of the total population, emphasizing that Hong Kong is still a society of primarily Chinese immigrants (Census and Statistics Department 2006, Kam-yee and Kim-ming 2006, Kim 1996). According to official records, the total number of foreign nationals rose from 134,000 in 1981 to 321,000 in 1993, and then to over 400,000 by March

---

6   It is important to note that before the handover of Hong Kong to China in 1997, Bangladeshi citizens could enter Hong Kong without obtaining a visa prior to their departure. They were allowed ingress to Hong Kong for three months upon arrival at the airport. After 1997, the immigration department set out revised entry visa/permit requirements for persons wishing to enter the Hong Kong Special Administrative Region (HKSAR). Accordingly, for Bangladeshis the duration of stay permit has been reduced to two weeks. Again, on 11 December 2006, the policy on entry permits changed. Under the new policy, Bangladeshis are not given any permit on arrival at the airport. They have to obtain a visa/permit before they fly. Having faced acute labour shortages, the Malaysian government entered into an agreement in the early 1990s with Bangladesh to recruit 50,000 workers every year. The government of Malaysia then placed a ban on the importation of Bangladeshi workers in 1996. In August 2006 Malaysia lifted the ban only to reinstate it, just two months later (*Malaysiakini* 2006). The current study was conducted between June 2004 and October 2006. Therefore, any changes in immigration or labour importation polices after December 2006 in either of the destination countries will not be dealt with or extensively covered by this research.

7   Of the rest, 5 percent is constituted by Pakistani, 3 percent by Nepalese, 4 percent by Thai, 5 percent by Indian, 6 percent by Japanese and Korean, 41 percent by Filipino, 15 percent by Indonesian, 13 percent by British and American, and 9 percent 'other' (Census and Statistics Department 2006 and 2003)

1995 (Cullinane 2003, Athukorala and Manning 1999). In 2004 it rose further, to reach 524,200 (*Asian Migrant* 2005).

Although in the early 1990s, the primary concern related to international migration in Hong Kong was with emigration and the possibility of 'brain drain', after the 1997 handover to China, the focus has shifted to immigration. Hong Kong has long had acute labour shortages, particularly between the late 1980s and early 1990s, in major employment sectors. To combat this, the government introduced the General Labour Import Scheme on the basis of the industry quota system which was later terminated in 1995, following the rise in unemployment and protests by local labour interest groups. Later, a Special Labour Importation Scheme (SLIS) was introduced to import foreign labour in order to complete specified major construction projects, such as the new airport (Kam-yee and Kim-ming 2006, Chiu 2001, Chan and Abdullah 1999). Thus the Supplementary Labour Scheme (SLS) was replaced by the SLIS (Kyokai 2003, Skeldon 1990).

Importantly, as reported by the 2001 census, over 40 percent of the Hong Kong population is now foreign-born (Hewison 2003). In 1997, foreign workers from Asian countries numbered 436,000, while there were approximately 121,000 from Western countries (Stahl 2003). Foreign domestic helpers (FDH, hereafter) dominate Southeast Asian migration to Hong Kong, and are mainly from the Philippines, Indonesia and Thailand (Ullah 2010, Li, Findlay and Jones 1998, Chowdhury et al. 2002, Law 2002, Asato 2004, Iyer et al. 2004). The introduction of FDH in Hong Kong dates back to 1974, however the number remained stable until the 1980s and 1990s. At the end of 2002, officially, the FDH population reached 237,104 (Asato 2004). Today, they constitute around one-fourth of the total population of migrant workers.

Highly skilled migrants to Hong Kong come mainly from key countries in the world economy and major investors, such as the United States and Japan. Unskilled and semi-skilled labourers unexceptionally come from less developed countries. The following figure demonstrates the striking lack of official data. The position of Bangladesh in terms of the volume of migrant workers in relation to the three major sending countries – the Philippines, Indonesia and Thailand – is the lowest. Although the number of Bangladeshi migrant workers in Hong Kong is small compared with these three countries, they have had an extended presence there, at least since the mid-1970s (BMET 2006). Even so, as a group, they have received very little research attention, with the result that we know little about this population of migrant workers in Hong Kong.

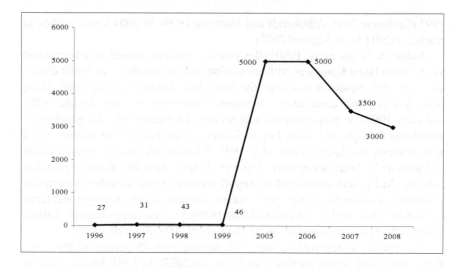

Figure 1.1 Presence of Bangladeshi migrants in Hong Kong
Note: From 1996 to 1999 the figure presents official data, after 1999 the figure presents the author's calculation (including students, FDW, labourers and businessman etc.).
Source: Asian Migrant Centre 2009, Hong Kong Immigration Department 2005.

Malaysia

Due to rapid industrialization, particularly in export-oriented industries, Malaysia in the late Nineties achieved a 2.6 percent unemployment rate which technically constitutes full employment (Athukorala and Manning 1999). This was accompanied by a massive population shift to urban areas as the Malaysian economy shifted from commodity production to manufacturing (Shari 2000, Pillai 1996). The economy has undergone dramatic growth and per-capita real domestic product more than doubled between 1950 and 1976 (Blau 1985 and 1986, Siwar and Mohd 1997). In addition, as with many of its Asian neighbours, Malaysia experienced a demographic transition from high to low fertility (Baydar, White, Charles and Ozer 1990), and a shift from a labour-surplus and low-wage economy to a labour-deficit and high-wage economy (Ariffin 2001, Skeldon 2000). Therefore, Malaysia began to experience difficulties with labour supply as its economy grew at 7.5 percent (on average) in the 1980s and 9.0 (on average) per annum[8] between 1990 and 1995, and the local labour supply increased at only 2.9 percent per annum over this period. GDP growth increased to 8.5 percent in 2004, compared to 5.8 in 1999. Kassim (2001) explains that the result

---

8   In 1980, the growth was 7.44 percent; in 1981, 6.94; in 1982, 5.94; in 1983, 6.25; in 1984, 6.70; in 1985, -0.96; in 1986, 1.05; in 1987, 5.39; in 1988, 8.94; in 1989, 9.21; in 1990, 9.74; in 1991, 8.66; in 1992, 7.79; in 1993, 8.34; in 1994, 8.75; and in 1995 it was 9.62 (IMF 1995).

was a labour shortage made worse by the increasingly selective attitudes of local labour as education and living standards improved, making Malaysian youths averse to accepting low status work (Abdul-Aziz 2001). Due to the labour shortages in the mid-1980s, many manufacturing units in Malaysia were in deep trouble, and even when the economy recovered from the 1997–98 economic crises, some companies were considering re-locating their manufacturing units to countries where labour was cheaper and more abundant (*Daily Star* 2002, IMF 2005) which provided fertile ground for the expansion of migrant labour. The year 2009 however saw a decline in the growth which was between 5 and 6 percent.

Foreign labour recruitment for economic development has become a significant element of Malaysian history. One policy solution for the problem of labour shortages was the importation of workers from Indonesia, Thailand and Bangladesh. Through various systems of recruitment, Malaysia imported labour migrants for its plantations, tin mines, and infrastructure projects, as well as the service sector. The sustained high economic growth Malaysia has been experiencing since its independence in 1957 brought about an increase in migrant workers in these particular sectors (Piper 2006, Athukorala and Menon 1996). Gurowitz (2000) points out that approximately three-fourths of the migrants in Malaysia come from Indonesia, with the balance coming from Bangladesh, the Philippines, Pakistan and Thailand. Officially, there were 2.1 million migrant workers in Malaysia in 2009 (Ministry of Human Resources 2009). However, Dannecker (2003) indicates — though debatably – that, at this time, the government of Malaysia was aiming to attract Bangladeshi workers in order to 'increase the number of Muslim migrants'.

In 1993, a total number of approximately 533,000 work permits were issued to import labourers to Malaysia, and this number increased steadily. According to Abdul-Aziz (2001), there were approximately 500,000 foreign workers in 1984, and their numbers shot up to beyond 1.5 million in 1991, and further to 2.4 million in early 1998. In 1998, 10 percent of Malaysia's population and about 27 percent of the country's labour-force were made up of foreigners, constituting the highest percentage of foreign workers in the world (*Migration News* 2005, Pillai 1998, Khan 2004). Precisely, in 1996, a total of 449,565 foreign workers were imported, of whom 71,254 were in domestic work, 204,614 in factory work, 75,944 in agriculture, and 10,841 in the service sector. By 1999, the number of migrant workers had gone up to 715,145, of whom 73 percent were from Indonesia, 19 percent from Bangladesh, and 3 percent from the Philippines. By 2002 it had risen again to 1,450,000 (1.45 million) which constitutes 12.8 percent of the total labour force of the country (Piper 2006). Again in 2004, it was estimated that there were 1,400,000 registered migrant workers and 1,200,000 illegal migrants in Malaysia. Malaysia had, in the first decade after independence, seen a gradual shift in the pattern of employment with a major shift from the plantation sector to the construction sector: by 2004, 414,300 migrant workers were employed in the manufacturing sector alone (Piper 2006).

Data further show that there were 81,000 Bangladeshi illegal migrants working in Malaysia in 2001, after which no official statistics are available. However, according to media sources, such as the *Daily Jugantor* (2004), the number was 40,000 in 2001. This discrepancy in the statistics is due to the lack of official data. It is believed that, owing to the presence of a large number of illegal migrants, it is difficult to obtain accurate data, though there are good estimates of global migration data from the International Organization for Migration (IOM) and the International Labour Organization (ILO).

The number of illegal migrants rose dramatically after the suspension of the labour-export agreement. In 1994, Malaysia signed an official agreement with the Bangladesh government to recruit 50,000 skilled and semi-skilled workers every year, including doctors and nurses (Rudnick 1995, BMET 2008, Abdul-Aziz 2001, Kibria 2004). Later, during the mid-1990s, however, this agreement was suspended by the Malaysian government. Many of the workers who had come officially extended their stay after their contracts expired, resulting in thousands of Bangladeshis continuing to work illegally or with forged travel documents (Christine 2003, Dannecker 2003, Netto 2001).

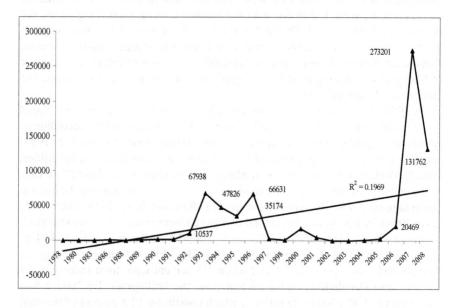

Figure 1.2 Growth of Bangladeshi migrants in Malaysia (1978–2008)
Source: BMET 2010.

As can be seen by the information above, Malaysia is a useful research site for investigating Bangladeshi migrant workers because of the remarkable volume of migrant workers in general, the long history of Bangladeshi migration to Malaysia, and the fluctuations of policy that impact the workers. To date, very little research attention has been given to the vulnerabilities of these migrant workers. As the Malaysian government allowed employers to recruit foreign labour, Bangladeshis moved into the paper mills, leather, rubber and garment factories, electronic companies, plantations, petrol stations, restaurants, and construction industries (Ullah 2010, Pillai 1998, Chantavanich and Risser 2000). The early 1990s saw a sharp increase in the number of Bangladeshi workers in Malaysia (Kassim 2001): in 1993, the largest number of Bangladeshi workers (67,938) was working legally there. Figure 1.2 illustrates the trends in Bangladeshi migration to Malaysia. The steep decline after 1997 was due to the suspension of the agreement (BMET 2010). The trend line with $R^2 = 0.1969$ presents a non-linear migration flow to Malaysia from Bangladesh.

Existing literature on Bangladeshi migrant workers

The body of literature on Bangladeshi migrant workers is evident in the diverse range of studies conducted in a multidisciplinary fashion. Migration as a field of study and research emerged in Bangladesh only during the early 1990s when a few organizations emerged with international sponsorship, particularly from the IOM. Most of their research has been clearly programme- and policy-oriented. Tasneem Siddiqui made commendable contributions to knowledge about migration as a discipline in Bangladesh. Her research covers key issues such as the effects of labour migration from Bangladesh on the trade unions (1999); the contribution of returnees (2000); migrant workers' remittances and micro-finance institutions (2001); the migration of female labourers from Bangladesh (2001); streamlining the labour recruitment process in Bangladesh (2002); migration as livelihood option for the poor (2003); the trafficking of Bangladeshi women and children (2003), working conditions for Bangladeshi factory workers in Middle Eastern countries (2004); and international labour migration integrating employment and working protection (2004).

Siddiqui's studies examine diverging aspects of the migration process, including factors inducing migration (push or pull); remittances, impact on development, the vulnerabilities of migrants, gendering migration, dimensions of migration flow, health risks associated with migration, trafficking in women and children, displacement, refugees and diasporic issues, and policy issues. Abrar (2000 and 2002) has also investigated the cost and benefit of migrants and refugees in Bangladesh. However, most of the studies clearly lacked theoretical and academic underpinnings.

Afsar (2003) has investigated the poverty-migration nexus in Bangladesh. She argues that poverty induces migration as much as migration contributes to easing poverty. The study found that issues of migration have remained missing in development policies. Many studies on Bangladeshi migration have linked migration with poverty, as did Skeldon (2004) in many of his studies. Afsar

(2001) in another study on 'globalization, international migration and the need for networking' illustrates that the communication system has been simplified due to the development of the information technology (IT) sector worldwide. The study, however, argues that only the privileged classes benefit from IT, and that only 0.02 percent of the total population of Bangladesh have access to such IT facilities. The study expressed concern about the derogatory attitudes, growing manifestations of racism and other forms of discrimination resulting from IT access.

While Rahman's studies on Bangladeshi migrants in the international labour market have covered a range of issues, including remittances and the role of networks, all of his work is concentrated on Bangladeshi migrants in Singapore. His studies on Bangladeshi labour migration to East and South-East Asia (2006), migration and poverty in Bangladesh (2004); Bangladeshi migrant workers in Singapore (2005), and Bangladeshi migration networks in Singapore (2004) suggest that, traditionally, labour migrants are both pushed by the lack of opportunities and poverty in their home country and pulled by the hope of economic gain in the destination country. His study on networking which was conducted on a particular village in Bangladesh showed that informal networking based on friendship and kinship ties is a central element facilitating migration to Singapore. Hassan (undated) in his study on 'complementarities between international migration and trade in Bangladesh' argues that international migration is a complement to commodity trade because migrants bring with them new skills and training, as well as cultural values.

Dannecker's (2005 and 1999) studies on 'transnational migration and the transformation of gender relations in Bangladeshi labour migrants' investigate economic factors for the increased migration flow of Bangladeshi women to Malaysia as temporary labour migrants and offer an analysis of networks and gender relations. Dannecker (2006) in another study entitled 'Bangladesh: Double standards' examines the role of female-based networks that have evolved to facilitate global mobility and a female-based credit system that has been developed by returned women migrants to give loans to other women to finance their migration. One of the most important works conducted by Dannecker (2003) entitled 'The construction of the myth of migration: Labour migration from Bangladesh to Malaysia', documents the various vulnerabilities suffered by migrant workers in Malaysian factories. Jones (1996) in his study 'Hope and tragedy for migrants in Malaysia', examines the development impact, and at the same time the vulnerabilities of the migrants to various exploitations.

Economic theorizing surrounding the household is couched in terms of decision-making but there is a divergence in the conceptualization of intra-household relations between those who deny that conflict is a factor in intra-household relations and those who allow for it (Kabeer 1997). The powerful place held by the male breadwinner in both development studies and policy studies has meant that the women's breadwinning role has been neglected (see Kabeer 2003). Most women worldwide engage in undocumented and often unremunerated work in the domestic, agriculture and industrial sectors (Kabeer 2004).

Oishi (2006) in her studies 'Women in motion: Globalization, state policies, and labour migration in Asia', and 'Gender and migration: An integrative approach' examines cross-national patterns of international female migration in Asia. Nana, in most of her studies, focuses on Asian perspectives with an emphasis on the gendered aspects of migration. In his research entitled 'Social capital or social closure and immigrant networks in the labour market' Waldinger (1997) examines the role of networks in providing efficient conduits for the flow of information and support. Spittel (1998) in his study on networking over-emphasizes the role of social networks, or 'migrant networks' in explaining international migration. In defining networks, his study focuses on interpersonal ties, reciprocal obligation and social connections. He applies inferential statistics to investigate the effect of networks on migration propensity.

Along with Castles, Massey and others, and Zeng (2004, 2003 and 1998, respectively) examine how Asian immigrants economically assimilate into the receiving society. The study argues that assimilation is an important factor in determining initial income as the study found that immigrants experience lower initial earnings. However, as soon as they are able to assimilate to the new society they tend to catch up with the income levels of native workers. Myles and Hou (2003) in their studies on spatial assimilation theory and new racial minority immigrants found that spatial assimilation is also an important part of the migration process. The International Organization for Migration (IOM), the International Labour Organization (ILO), and INSTRAW have commissioned several studies on Bangladeshi migrants based on empirical data. Abdul-Aziz (2001) conducted one of these studies on Bangladeshi construction workers in Malaysia that focuses on skill composition and wages as well as, to some extent, migrant vulnerabilities.

The current literature, mentioned above, on Bangladeshi labour migration has clearly demonstrated a specific knowledge gap in migration studies as, evidently, the researchers have hardly focused on issues of rationalization of migration decisions. This study, with the objective of understanding and investigating the rationalization of migration decision for Bangladeshi migrants in Hong Kong and Malaysia, addresses the problems and vulnerabilities faced, migration costs, access to jobs, income and savings, and remittance transfers as well as its uses at home and its ultimate impact. In covering this information, the current study seeks to fill in several significant gaps in the migration scholarship in terms of the situation in Bangladesh. While this research does not claim to offer a complete analysis of rationalization of migrants migration decisions, given the limited sample size and scarcity of official data, this study nevertheless intends to substantiate the existing research on labour migration in the two cases under study and extend theoretical explanations of rationalization addressing the current migration trajectories that have remained untested.

This book does not intend to present the factors influencing migration; rather it investigates how migrant workers undertake a logical reasoning for decisions already made. In their discourse, they offer logical reasons (as will be demonstrated in the subsequent chapters) that they use to rationalize their decisions to migrate.

It is a challenging task to analyze and measure 'rationalization' primarily because there are many unseen and unpredictable factors associated with this process, as well as its subjectivity. In addition, rationalization does not simply imply a state of mind that prompted potential migrants to make a decision to migrate; rather it entails many other factors, in particular post-migration experiences, which are not often existent, and therefore not taken into consideration, during the decision-making process. Subsequently, migrant workers suffer frustrations in the long run as they face various difficult realities after migration has occurred.

Despite the long history of massive migration of Bangladeshis for employment, interest in studying migrant workers has grown only recently. The recognition of the significance of migration can help to ease Bangladesh's unemployment crisis and contribute to economic development: Therefore, it warrants deeper analysis and understanding. For this reason, the process of how migrants rationalize their migration decisions is crucial, as it relates to how they communicate about their migration experiences to others, potentially inducing them to migrate. This is particularly the case with under-educated migrant workers, who are especially vulnerable to a wide range of exploitations, some of which may devastate them financially and cause further economic harm.

The study of rationalizing migration decisions encompasses issues of both pre- and post-migration periods, including the cost of migration, future income benefits, and possible adversities and risks. Decisions are generally based on expectations and promises of better jobs, economic gain, and subsequently, better future prospects. However, with a few exceptions, migrant workers often experience under-payment, challenging relationships with employers, and in some cases, confinement to the work place, withheld and delayed or irregular salary payments and confiscation of travel documents. Therefore, a statement of the study's research argument is formulated as follows: Bangladeshi potential migrant workers from diverse socio-economic backgrounds decide to migrate overseas with a set of expectations in mind, however, in most cases, their expectations remain unattained and therefore they necessarily tend to rationalize their migration decision in order to protect their mental well-being.

This book aims at contributing to an understanding of contemporary studies on migration, and looks into two periods of and perspectives on the migration process: one is the pre-migration period and decision-making perspective, and the other is the post-migration period and rationalization perspective. This study specifically seeks:

- To understand the rationalization of migration decisions and subsequent experiences during the post-migration period of Bangladeshi migrant workers in Hong Kong and Malaysia;
- To identify the problems and vulnerabilities faced, migration costs, access to jobs, income and savings, and remittance transfers, pattern of expenditure by the family at home, and the ultimate financial impact of the migration;

- To recommend a set of policies and strategies that will enhance the benefits of migration, both for the migrant workers themselves and for the governments at both ends.

Plan of the book

This book is composed of eight chapters. Each of the chapters – logically arranged according to the theoretical framework – answers the questions the book poses. While the variables influencing migration could be different, the book focuses on the rationalization process post-decision-making and the migrant workers' post-migration experiences.

The second chapter introduces the major theories of migration and presents the framework of the research the book deals with. This chapter – divided into two major sections – presents information on research and sampling design as well as details on the data collection and analysis. One section deals with the theoretical perspectives while the other deals with the methodology. One of the main objectives of this chapter is to present several dominant theories of migration and to develop a theoretical framework. This chapter clearly sets out the two perspectives on the migration process (pre-migration decision-making and post-migration rationalization) and demonstrates their association with the selected theories. It also indicates how the following chapters link with the theories to achieve the objectives of the research and address the argument. The second objective of this chapter is to present a systematic account of the methodological issues, stating the entire process, from the research plan to its operationalization.

The third chapter focuses on the demographic profile of the respondents and provides a brief socio-economic and geo-demographic background for both interviewed groups – Bangladeshi respondents in Hong Kong (BRHKs) and Bangladeshi respondents in Malaysia (BRMs). This chapter also examines some characteristics of the family and how they may or may not have influenced migration. The two investigated destinations are heterogeneous in terms of economy. The chapter indicates that the respondents came from both a socio-economically and demographically homogenous background with no significant differences.

The fourth chapter highlights the routes through which the migrant workers travel to get to their destination countries, and also investigates how networks facilitate migration and how different social groups become involved in the process. Theoretically, migration networks are significant actors in both pre- and post-migration periods, in terms of reducing the risk and cost of migration; therefore they have a significant influence on the migration decision. This chapter further demonstrates different levels of network involvement and how they are operative in inducing migration.

The fifth chapter deals with one of the most important issues of migration: Migration cost and ways of financing initial migration. Although migration costs are a well-addressed issue, this chapter includes several factors in its analysis which are less commonly addressed; for example, the sources for borrowed money, necessary to finance migration, as well as how long it took for migrants to repay their loans. Therefore, not only the monetary cost is analyzed but also the social and psychological costs of migration debt.

This sixth chapter discusses the influence of work-related issues – the categories of work offered/available to migrants, contracts, employee-employer relations, and wages – on the rationalization process. This chapter further analyzes the adaptation of migrant workers to their host societies, which is a significant element in the rationalization of their post-migration experiences. This chapter demarcates two significant post-arrival aspects: living conditions and the resulting stressors, as well as frustration due to a variety of job-related issues.

The seventh chapter deals with the process of deriving and transferring remittance. Furthermore, this chapter identifies the channels of transfer, the dynamics of remittance use and its impact on the well-being of the receiving families, and also examines the role and influence of income in the host countries on migrants' rationalization processes. This chapter deals with two significant variables in the rationalization of migration decisions: income and remittance. This includes the level of income earned and various aspects of remittances, such as how they are constituted and what the modes for transferring are and why. However, determining an empirical relationship between remittance and economic growth is complicated by problems of endogeneity, associated with difficulties in finding adequate instruments to explain the behaviour of remittance. The findings in this chapter substantiate the assertion that a large amount of remittances are being transferred to Bangladesh through informal channels. Due to the illegal transfer of a large amount in remittances, the ultimate looser is the Government. Therefore, the significant flow of remittances into the Bangladeshi economy annually has generated considerable interest in the country's development discourse in recent years.

The eighth chapter synthesizes all of the preceding chapters and provides concluding remarks and recommendations as well as some implications of the study. It assesses each of the chapters in relation to its objectives and research questions.

Chapter 2
Theoretical and Methodological Considerations

This chapter discusses and critiques several of the main theories of migration studies and presents a theoretical framework explaining two perspectives (pre-migration decision-making and post-migration rationalization and experiences) of the migration process. The first section of this chapter provides the theoretical framework to interpret the research. The second section offers a detailed methodological guideline for the collection and analysis of the current study's empirical data.

Rationalization: Theoretical perspectives

Before discussing specific theories, I would like to introduce the concept of rationality and rationalization, and how rationalization is conceptualized in this particular research as it is central to the theoretical hypothesis. Before discussing the main theories, it is also important to note how rationalization is conceptualized from psychological, sociological and economic perspectives. Decision-making and rationalization of any action are based on the idea of rational choice[1] which takes into account not only preferences and expectations but also the opportunity structure of the context encountered by the actors (Esser 2003). The more powerful the motives of a migrant to achieve a particular goal the stronger are his or her rationalized expectations that this goal can be achieved.

Rationality, both linguistically and conceptually related to rationalization, refers to the possibility of goal attainment. Therefore, an act or decision is rational when it is the optimal choice in pursuit of a goal. Weber suggested four modes of rationality: formal/purposive/instrumental rationality; substantive/value/belief-oriented rationality; affectual rationality, and traditional rationality (Cockerham, Abel and Lucshen 1993).[2] According to formal/purposive/instrumental rationality,

---

1 Rational choice theories hold that individuals must anticipate the outcomes of alternative courses of action and calculate which will ultimately be the best. Rational individuals choose the alternative that is likely to give them the greatest satisfaction (Lee 1966, Heath 1976, Carling 2005a, Coleman 1973).

2 Substantive/Value/Belief-oriented rationality relates to the action undertaken for reasons intrinsic to the actor: whether ethical, aesthetic, religious or another motive, independent of whether it will objectively lead to success. Decision making of this kind is

decision making is based on calculation, efficiency, and predictability, technical appropriateness, and orientation towards personal or institutional utility. Weber began his studies of rationalization with the Protestant Ethic and the Spirit of Capitalism, in which he illustrates rational means of economic gain as a way of dealing with uncertainty. Habermas (1984) has argued that to understand rationalization properly requires going beyond Weber's notion of rationalization and distinguishing between instrumental rationality, which involves calculation, and efficiency and substantive rationality. Alternatively, March and Olsen (2004) termed rationalization as the logic of appropriateness which means that human beings play role when they think it is right or appropriate. Therefore, action is driven by rules of appropriate behaviour.

According to the Weberian concept of rationalization, an action is considered rational if it leads to the achievement of a goal for the lowest cost, time, effort, resources, and level of difficulty (Carl and Fetzer 2005). The lower the cost the easier the rationalization of the act becomes. This raises the question as to what rationalization of any action entails since different phenomena involve different variables. The rationalization of migration implies that decisions to migrate are made correctly or incorrectly according to the perceived loss, benefit, employment prospects, and vulnerabilities involved (Habermas 1984). Migration decision-making at the individual level, therefore, is a process of identifying and choosing the best option from available alternatives based on one's individual values and preferences (Sadler 1989). These may be related to improving and securing wealth or status, an increased level of autonomy, personal affiliations, or providing a more meaningful and better life (Rahn, Krosnick and Breuning 1994).

Sociologically, rationalization is the justification of an action performed. From the psychological point of view, *rationalization* is the process of constructing a justification for a decision that was originally arrived at through a different, often less concrete, mental process such as opinion. Rationalization occurs because human beings often feel the need to construct a logical reason for an action of which the 'moral superego disapproves ... the ego seeks to defend itself by adding reasons that make the action acceptable to the superego'. Furthermore, rationalization acts as a defence mechanism that justifies an action or an evaluation made by beliefs (Parsons and Edwards 2001). Rationalization as a defence mechanism is significant in this research because the current study analyzes respondents' level of satisfaction or dissatisfaction with various aspects of their migration decision.

---

based on ultimate values whose significance and worth transcends personal or institutional utility. Affectual rationality is determined by an actor's specific affect, feeling, or emotion – of which Weber himself said was on the borderline of what he considered 'meaningfully oriented'. Traditional rationality is determined by ingrained habituation. Weber emphasized that these are ideal types of rationality, and it is unusual to find only one of these orientations used in any decision making process: combinations are the norm. His writings also make clear that he considered the first two forms of rationality as more significant than the others, and it has been argued that the third and fourth are subtypes of the first two.

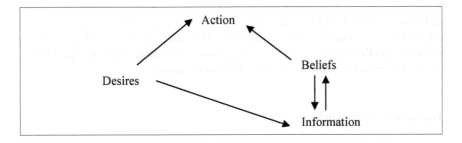

Figure 2.1 Rationalization model
Source: Elster 2000.

In compliance with the above, Elster (2000) indicates that an action should be the best way of satisfying a set of desires given an individual's beliefs. Therefore, a rational action (decision or choice) is one that is both caused by the desires and beliefs of an individual and optimal in the light of those desires and beliefs (Elster 2000). Despite such rationalization, a calculated action may result in either negative or positive outcomes.

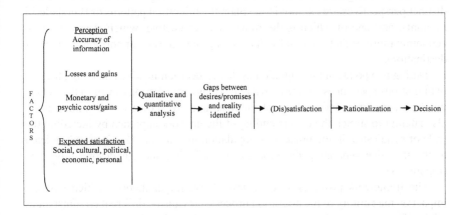

Figure 2.2 Model of rationalizing migration

In view of the above, sociologically, psychologically and economically, people's rationalization of migration illustrates how migrant workers justify their acts of migration through their beliefs and choices made according to certain patterns. Figure 2.2 shows the migrant workers' rationalizing process and their resultant migration decisions. In this book, the analysis of the process of rationalization of migration decisions advances according to the proposed model. The model clearly demonstrates that the perceived losses and gains, the cost paid (monetary and non-monetary), the level of satisfaction for various factors (income, living condition, socio-political status), vulnerabilities and adaptation strategies, when analyzed quantitatively and qualitatively, identify the gaps between expectations for migration and the realities of post-migration experiences.

Theoretical framework

I will discuss the individual theories which have formed the basis of the framework for this book (Figure 2.3). Massey et al. (1996) suggest that research on patterns of international migration needs to be situated within a well-articulated framework in a way that links explanations of migration processes and outcomes at various conceptual scales. Therefore, the theoretical framework has several implications for the analysis of migration studies (Barnett 1999).

To provide a theoretical framework which is coherent with the current study and corresponds to the research objectives, the most relevant theories have been chosen in order to place the migration phenomenon in question (i.e., the rationalization of the decision to migrate) under scrutiny on both an empirical and theoretical level. The framework explains the rationalization experiences and decision-making processes of Bangladeshi migrant workers in post-migration situations. Some distinct perspectives are explored to explain the migration patterns of Bangladeshi migrants, and one of which is the functional perspective, which is premised on economic motives that drive individuals and groups to seek opportunities in other destinations.

During two points in the migration cycle i.e., decision-making and rationalization of migration experiences, two principal issues play a role. Various levels of theoretical involvement correspond to these phenomena and are shown in the framework. I also attempt to stretch the understanding of the network approach by including the 'role of syndicates' in the process of population movement. This is closely related to the thin line separating the process of trafficking and clandestine migration (Figure 2.3).

The framework provides an overview of five major theories which have an impact on the entirety of the migration process. Obviously, different circumstances elicit different affects. However, economic considerations have always registered a high level of influence throughout the migration process. While networking and household strategy play a stronger role at the decision-making stage, economic and adaptational factors have are powerful influences for both rationalization and decision-making.

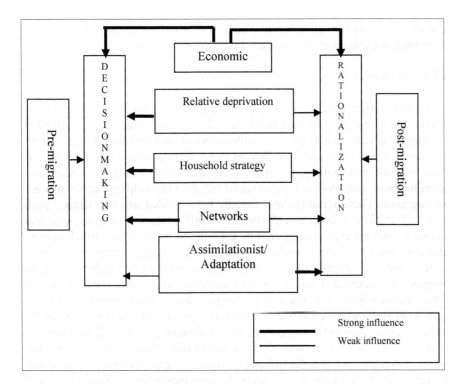

Figure 2.3 Theoretical framework

Dominant theories of migration research

This section discusses different approaches and theories in the study of migration and provides an explanation of the proposed theoretical framework which is applicable to the context of Bangladeshi migrant workers. Social scientists have offered a number of theories to explain the changing patterns of migration leading to approaches with different levels of analysis (for example, micro versus macro). The phenomenon of migration cannot be studied in a mono-disciplinary fashion as migration embraces a range of perspectives covering the sociological, economic, political, cultural and global realms. It seems reasonable, therefore, to analyze migratory patterns by importing notions and views from a diverse range of social science disciplines (DeFay 2004, Brettell and Hollifield 2000 and 2000a). Migration research, as a discipline, is over 100 years old and most extant studies are inquiries into the factors that induce people to migrate. Although studies on migration have been well covered by this research, several important questions have been left open, in particular: Why do people move? (Ainsaar 2004). After

Ravenstein's (1885) migration rules,[3] tens of different migration indicators have been used as explanatory factors, each involving different coverages.

Classical theories

In analyzing migratory patterns, migration theories generally revolve around three central questions: Why do people migrate; Why do they stay in the destination country; Why many people do *not* consider migrating despite conducive conditions outside their place of origin (Nikolinakos 1996). Efforts to answer these questions have resulted in the emergence of the classical theories and these studies have identified economic, political and/or social variables that help explain the process of migration. These variables have broadly been divided into two factors: push and pull (Chiuri, Giuseppe and Giovanni 2005). Classical and neo-classical economic models explain migration by referring to these factors, which assume that two forces i.e., push to emigrate in the country of origin and pull immigrate in the country of destination, contribute to the conditions that induce people to migrate. As indicated above, the pattern, composition, dynamics and volume of international migration is constantly changing due to displacements and other global circumstances such as war, economic collapse and famine (Lewellen 2002). However, migration in the current age of globalization presents different patterns from earlier examples (Shuval 2000). A contributing factor is likely to be that social inequalities between the rich and the poor have grown exponentially, in part as a result of globalization, which essentially promotes the emergence of push-pull conditions (Anthias 2000, Yang 2003).

Apart from the push-pull forces, with better communication ability, easier and cheaper transportation systems, and the effects of liberal trade and international

---

3   A little description surrounding the gravity approach (GA) is required here, as it has been one of the most influential theories in migration study since Ravenstein's time, and hence I must provide logical ground for excluding it from my model. The Gravity Approach stipulates that there is a relationship relation between distance and the propensity to move; however this factor has very little influence my on the current study. For example, one of the GA propositions is that the majority of migrants move only short distances; however, as Massey et al. (1996) assert in the world system theory (WST), countries need not be geographically close for migration in significant numbers to take effect. Currently, Bangladeshi migrants set off for distant countries such as Australia, Canada and even areas of Africa in order to seek economic improvement (Ossman 2004, Palloni 2001, Ullah 2005). It appears that in these migration decisions, the distance factor does not play a major role. According to the basic instrumental statement in the value expectancy (VE) and structural individualism (SI) models, migrants decide between at least two alternative courses of action, and in so doing, chose the one for which the perceived value of the result is greater. Therefore, the actor is able to make a rational decision on the basis of a set of value or preference orderings (Dejong and Fawcett 1981). These may be related to improving and securing wealth, status, comfort, affiliation, or exit from professions of all kinds, and a meaningful life.

capital-flow policies, the current world economy has been experiencing more efficient transactions between regions than ever before (Davies 2003). According to Cohen (1996), international migration is, therefore, affected less by wage or employment differentials between countries, than by policies towards overseas investments and the international flow of capital and goods (Sassen 1998, Potter 2004). Thus, most migration theories analyze the conditions which initiate the international movement of labour, for example wage differentials in neoclassical theory and labour-demand-based pull factors within the dual labour market theory including the spread of migration networks and other institutions supporting or repelling migrants (Massey et al. 1999).

In addition, according to many empirical studies, poverty is a major cause of both national and international migration flows, as Skeldon (2002) has illustrated in his strong correlation between poverty and migration.[4] Ullah and Routray (2003 and 2007) demonstrate similar findings, however, they have observed a more nuanced pattern in that, while most women in Bangladesh experience poverty, their propensity to migrate is the lowest when compared with those of neighbouring Asian countries. At the outset, this may seem to be a case against the poverty-driven migration theory. However, the situation of Bangladeshi women is at least partially explained by rigid social and religious sanctions that restrict their movement overseas.

Most researchers on international migration who have dealt with the factors that prompt migration itself suggest that multiple factors have an impact on people's decisions to migrate, and affect peoples' decisions in a variety of ways. This section touches on the focus of the current study which inquires as to how migrant labourers rationalize their migration experiences. However, my current research will investigate the question of how people go about their decision-making processes as well as how and what rationalization (post-migration) issues they grapple with, which have not been adequately addressed in previous studies. Therefore, this research examines Bangladeshi migrants' rationalization of their decisions at two points of time during the migration process:[5] the first is how the migrant workers rationalize their migration decision pre-migration; and the second is how they rationalize their current situation (post-migration rationalization).

Economic approach   This research is influenced by different economic models. While income at the latter stages post-migration determines the length of residence for migrant workers abroad, the cost for the financing of migration influences the initial pre-migration decision-making process. This indicates that financial matters

---

4   The human development gap is aggravated by substantial gender disparities with a difference between HDI rank and GDI rank of approximately -5 (Neft and Levine 1997). Women have been hit by financial, economic, and social crises in varying degrees worldwide; and religion, in addition, continues to restrict women's interests in many countries.

5   Whole cycle refers to the cycle of migration process i.e. 'from pre-departure – post-arrival, stay and return home'.

are a determining factor in how migrant workers rationalize their migration experiences. Financial factors considered include, but are not limited to, concerns related to obtaining the required amount to cover migration costs (the availability of capital), and the terms and conditions of repayment. Therefore, economic factors are likely to be one of the most significant components in migrants' rationalization and decision-making processes.

As an all-encompassing approach, including economic issues in my analysis allows the inclusion of related ideas like cost-benefit, human capital and dual labour market theories. The central argument here is that migrants relocate internationally primarily due to economic opportunities (Chiswick 2000 in Brettel and Hollifield 2000). This concept echoes the fundamental ideas of classical and neo-classical economic theories, according to which economic factors function to push and pull people from one place to another. According to Ranis and Fei (1961, Skeldon 2003) migration is a response to a high demand for labour by an industrial sector which assures workers greater levels of productivity, and investors positive profits superior to the opportunities found in the traditional agricultural sector. This reflects that demand factor which acts as a pull force for a potential migrant from a labour abundant country with a dominant agricultural sector.

Cost-benefit model   In the economic approach, the idea of costs-benefits is one of the potential components. The costs of migration remain myriad and vary across places and persons, changing rapidly over short intervals of time. The 'cost' of migration may include direct financial costs, indirect opportunity costs and social or emotional costs (Zeitlyn 2006). According to this approach all human beings are treated purely as economic actors, who decide on their move after an analysis of the possible costs versus the possible benefits and the likelihood of each. The cost-benefit approach (CBA) has been performed by Chiuri, Giuseppe and Giovanni (2005). They theorize that migrants make their final decision to migrate when the calculation of the CBA results in perceived future monetary benefits (i.e., post-migration benefits) that are higher than the sum of the present value of income and the total monetary cost of moving (Abrar 2002, Miyan 2003). Therefore, in this model, monetary benefit is the primary criterion that pulls migrants from one place to another. If we were to apply this approach to the current study, it would appear that Bangladeshi migrant workers in Hong Kong and Malaysia have moved primarily due to economic factors, either acting as push forces, i.e., economic hardship in the country of origin, or pull forces, i.e., perceived economic opportunities in the country of destination.

However, economic indicators might not always provide sufficient explanation for the international migration of people from Asia into the global labour market (Abdul-Aziz 2001). Within an individual rational choice model aimed at long-term income maximization, the potential migrant calculates costs and benefits, and expects a positive net return from his or her decision based on the available opportunities to capture a higher wage rate associated with greater labour productivity. The underlying assumption of the economic cost-benefit model is generally expressed by equations.

There is no doubt that monetary reward constitutes an important factor that influences a prospective migrant's decision. According to the CBA, a potential migrant is likely to move if the present calculated value of all future monetary benefits from moving is greater than the monetary cost of moving, including the income differential. However, it is often impossible to calculate the psychological, social and opportunity cost involved. The proposed theoretical model thus includes one factor at the emigration country, and one factor at the immigration country. Therefore, the major drawback of this model is that the psychological and social costs to the migrants and the families they have left behind are not included due to the difficulty in quantifying this variable. In addition, there is no single, clear direction to calculate the opportunity costs of migration: In calculating future benefits, risks, uncertainties and possible loss of work, among others, could not be accuragely predicted or considered. The major problem of the cost-benefit analysis is that it is exclusively a quantitative tool, and as such, cannot be applied to measure the many qualitative aspects of the migration process. Dannecker (2003) makes a good point that there are some aspiring migrants who set off for overseas simply out of a desire to travel on an airplane and to experience the lifestyle of foreign countries, as expressed by a migrant worker from Bangladesh (*JaiJaidin* 2006). Political persecution, and therefore personal safety, may also elicit a migration decision. Therefore, it cannot be said that only monetary considerations play a role in decision-making for migrants.

Neo-classical economic macro-theory

This theory attributes patterns of international migration to the demand and supply of labour, in coordination with wage differentials. Many argue that GDP and unemployment rates do not explain the patterns of international migration (Massey et al. 1996). Rather, they view geographical variations in the supply of and demand for labour in the origin and destination countries as the central factor that drives individual migration decisions. The assumption of this model is that international migration does not occur in the absence of these differentials and the elimination of the differential would bring an end to international labour movement. Furthermore, this theory holds that only labour markets act as the primary mechanisms for inducing international labour movement (Roy 1999, Massey 1996).

Like neo-classical economic theory, dual labour market theory highlights that international migration is demand-based and initiated by the recruitment policies of employers or governments in destination countries. Piore (1979) argues that international migration is caused by a built-in demand for immigrant labour that is inherent to the economic structure of developed nations. Immigration is therefore not caused by push factors in the sending countries, but by pull factors in the receiving countries. Keynes (1935) revised this model when the continued decline of the depressed industrial economy of Britain caused doubts about the 'capacity of labour mobility to produce equilibrium particularly when it was noted that those tending to leave a depressed region were often the most highly skilled' (Jackson 1986). However, empirical research and theories recognize that migration does not occur solely due to a single factor: There might be sufficient global pull forces, eliciting push forces to stimulate potential migrants to look for opportunities and benefits abroad. In addition, human capital has a significant function in galvanizing migration intention (Guerrero and Bolay 2005, Dudley 2002).

The neo-classical economic model places emphasis on the demand or pull factors that induce migration, although it is, in itself, not a sufficient condition or sole factor in drawing people away from their places of origin. Factors such as rising unemployment in countries of origin also play a significant role in driving people to seek greater opportunity elsewhere, a finding confirmed by multiple studies (Siddiqui and Abrar 2003, Maharaj 2004). War, widespread poverty and religious and political persecution often intermingle with and blur demand forces. Despite these drawbacks, the neo-classical economic theory is important to the current research because it emphasizes the push factors which are the clearest explanatory factors expressed by Bangladeshi migrants in making their migration decisions.

Nikolinakos (1996) emphasizes that economic theories do not include political factors, which influence economic processes (Maharaj 2004). He also considers the social relationships within the framework of which economic phenomena occur. Finally, Nikolinakos has argued that economic theories overlook demographic factors and their relationship to given economic conditions (Nikolinakos 1996). Moreover, micro-economic models of migration typically disregard the role of the state in mandating, proscribing, and regulating movement across national and international boundaries. Although there has been an outpouring of work on the politics of international migration over the last decade, it is characterized by 'thick description and inference from case studies that seem unlikely to lead to a general theory' (Zolberg 1998 and 1999, Hollifield 2000). Macro economic theory, neo-classical and dual labour market theories are rigid with respect to purely economic demand factors and condone wage differentials and unemployment-driven push factors. The current study takes into consideration that wage differentials in many ways determine the country of destination; therefore influencing a migrant's decision (indeed, many migrants rationalize their stay using the idea that they cannot earn at home as much as they do abroad with the same amount of time and labour), however, other factors play a part.

Human capital   This section discusses how significantly human capital (HC) matters in migration decisions. The theory of human capital is very important to the current study. The HC model takes the following elements into account: age group, education, skill composition and gender (Faist 2000, Ainsaar 2004). MacPherson and Gushulak (2004) defined human capital as a commodity to be recruited for the purposes of the labour market. The basic idea of the HC model is to regard migration as an investment that entails the cost 'now' in the hope of recouping future benefits. In its simplest version, a potential migrant calculates the discounted future income stream minus costs for each region, and then moves to the region with the largest discounted benefits[6] (Berninghaus and Seifert-Vogt 1992, Zimmermann 1992).

A closer look at empirical studies reflects the importance of human capital in migration trends. For instance, several studies confirm that relatively younger populations are more likely to migrate than older groups who stay home to take care of the family. Studies have also emphasized the value of skill-formation as a factor in migration decisions. However, I argue that, while skill-formation is an important variable in the decision to migrate, education level does not always play a role in the case of unskilled labour migration because education is normally not a pre-condition for this group. Two main explanations are proposed from an economic point of view: Younger populations usually have better labour market prospects in the receiving country because, in most cases, they are willing to take lower-paying jobs. Secondly, according to human capital theory, younger people tend to have better returns on investment when it comes to migration. Older people usually have fewer economic incentives to migrate (Bauer and Zimmermann 1999). Therefore, I agree with Krieger (2004) that younger groups consequently develop a greater willingness to migrate while older family members tend to remain in the country of origin in order to maintain the family (Krieger 2004). In addition, socio-economic models suggest that younger people have a higher degree of dissatisfaction with existing conditions in their country of origin. Combined with higher levels of aspiration, this often results in strong feelings of relative deprivation and frustration which prompt the potential migrant to decide to migrate (Stark and Taylor 1991).

Further, in studying the issue of selectivity in migration, Borjas (1994) presents the human capital model as an alternative specification of the Roy model, which implicitly assumes that there are no fixed costs and that ability has no effect on efficiency in migration. As a result, migration incentives are a function of wage ratios between the destination and origin regions (Roy 1999). However, displacement, occurring either voluntarily or under pressure, may be rooted in economic disparity which results in the feeling of relative deprivation.

Relative deprivation While labour demand in industrialized countries acts as a pull force to attract potential migrant workers, relative deprivation functions as a push force, enticing individuals to move in search of better opportunities. Migration, therefore, takes place as people who live in economically distressed areas feel 'relatively deprived' compared to other people within or outside their

---

6   However, Carling (2005) is concerned about the loss of human capital – 'brain drain' – which occurs in the migrant-sending countries. This is explained from the perspective of the state in that the emigration of skilled individuals represents a loss in its own right, and that the returns on government investment in education are lost to another country.

own communities. Hence, migration stems from people's desire to improve their relative position on the socio-economic ladder (Faini and Venturini 2001, Ghatak 1995). As long as high levels of disparity and widespread poverty prevail in much of the Global South, the attraction to migrate to more developed countries will continue to grow (Maharaj 2004). According to Starks' portfolio investment model, the twin factors of 'relative deprivation' at the place of origin and the desire for 'improving [one's] position' are the principal forces that drive potential migrants to move. Previous research by the current author confirms that, in search of improvement for their overall well-being, people are motivated by the desire, in some cases, desperation, to set off in search of overseas opportunities (Ullah 2005).

Stark's (1994) emphasis on the portfolio investment model shifts the focus of understanding migration decisions from an individual independence (as in the Harris-Todaro model) (Todaro 1969 and 1985, Harris 1970) point of view to that of interdependence. Remittances from migrants to their families at home, as well as a number of overt and covert interfamilial exchanges, are results of a collective migration decision. This model also argues that migration in the absence of a significant wage gap between the developed and underdeveloped regions, or the lack of migration in the face of a substantial wage gap, does not imply irrationality. Decisions to migrate may depend on wage uncertainties and relative deprivation at home, which could force families to pool risks and alter the pattern of human capital investments (Chiswick 2000). According to the economic theory, while economic factors shape the decision-making process of the migrants, there are also allied factors which are important in grasping a network approach, such as the destinations chosen, how migration is managed, and the cost of financing migration.

Several recent studies suggest that 'relative deprivation' induced potential migrants often make the decision not only to migrate, but also to remain in countries of destination (*Shaptahik2000* 2005, Siddiqui and Abrar 2003). This, however, does not fully explain migration trends if it is assumed that poverty is the only cause of emigration (Hye 1996). Therefore, it is unrealistic to assume that economic motivations are the sole reasons for labour movements, as these approaches have argued. The desire to see other countries, an unfulfilled love or marriage, or the political situation in the country of origin also contribute to migration decision-making. However, as argued by Dannecker (2003), the crossing of borders in a physical sense is closely connected with the hope of economic improvement.

Household strategy   The fact is that, in Bangladeshi society, a family is economically dependent on the household head, most commonly elder adult males (GoB 2005). Their absence affects the entire family emotionally, financially and socially. In addition, the amount of money required to finance overseas migration is normally too high to be easily obtained through a single source (e.g., selling property) and therefore must be collected through multiple sources, often involving borrowing. The sale of valuable assets and/or the accumulation of debt affects the family and their shared economy as a whole. Therefore, viewing migration decisions as household strategies i.e., the proposition of a collective economic strategy, is vital for understanding the decision-making process, and therefore occupies a significant space in this book.

The household strategy model presupposes that people act collectively in order to maximize expected income and to minimize risks for the members of the kinship unit. Therefore, the household is significant in controlling economic risk by diversifying the allocation of resources (Lieby and Stark 1988, Oishi 2002). This paradigm rejects that migration decisions are made at the micro level i.e., only by individuals. The household strategy likewise explains international migration as a means of compensation for the absence or failure of certain types of markets in developing countries, such as unemployment insurance. In contrast to the neoclassical model, wage differentials are not seen as a necessary condition for international migration, and economic development in areas of origin or equalization of wage differentials will not necessarily reduce the pressure to migrate. Massey et al. (1996) have supported this in the following statement:

> In developed countries, risks to household income are generally minimized through private insurance markets or governmental programmes, but in developing countries, these institutional mechanisms for managing risk are imperfect, absent, or inaccessible to poor families, giving them incentives to diversify risks through migration. Household strategy theory asserts that migration is a family (i.e., group) strategy to diversify sources of income, minimize risks to the household, and overcome barriers to credit and capital. (Massey et al. 1996: 186)

As previously stated, within households there are always some members who are able to make decisions independently by virtue of their age or financial strength or control within the family. Here lies the central weakness behind the household strategy. In addition, Dannecker (2003) pointed out the feminist critique of the concept of households, in that they are not homogenous units but, rather, stratified by gender (as will be demonstrated in the following section) and generation with conflated interests, embedded in a variety of networks. In the case of male Bangladeshi migrants, it can be argued that power relations between the different generations are one of the reasons why young men are choosing to migrate. Some migrants have very little say in the negotiation processes concerning possible migration (Goldscheider 1996). The possibility to escape from their families and the embodied power structures, including familial responsibilities, which are a component of their position within the given household setting, may contribute to their desire to leave.

In Bangladeshi society, households are seen as the primary unit for the migrant, which they refer back to. Households are based on a set of rules that are preset and are locked in practice due to the perceptions and traditions of generations (Karn 2006). In a household all the individual members share the same cultural reference point and they divide roles and responsibilities among themselves (Banerji 1993,

Werner 1998), often determined by age and gender hierarchies. The eldest male generally tends to make decisions for the joint family; female members generally do not play role in family decision-making. This position passed on to the eldest son when he becomes economically more powerful in the household (Karn 2006, Tientrakul 2003). Therefore, household members do not hold equal power. Hence the general idea of household strategy that the decision to migrate is made 'collectively' is not perfectly accurate, as all family members do not hold equal roles in the decision making process.

Modern theory

Network approach A network, in general, is a series of points or nodes interconnected by communication paths. In the migration context, networks are significant factors in shaping decisions to migrate. While classical models stress push or pull factors as driving forces that influence migration decisions, networks shape the destinations, costs and risks of migration. Migration networks are normally based on informal networking which is limited to friends and relatives involved in facilitating migration (Rahman 2003). Dannecker (2003), referring to Lomnitz (1977) and Massey et al. (1989 and 1988) demonstrated the relevance of networks to migration decisions and the direction of migration flows, illustrating that networks are sets of relations that connect migrants, former migrants, and non-migrants both in origin and destination regions through ties of kinship and friendship (Massey et al. 1987, Boyd 1989, Dannecker 2003). In receiving countries, immigrant communities often help their countrymen/women and kin to immigrate, find jobs, and adjust to the new environment (Kritz, Lim and Zlotnik 1992, Portes and Sensenbrenner 1993). In further explaining networks, Waldinger (1997) suggests that they, in fact, provide the mechanisms for connecting a highly selective group of 'seedbed' immigrants with a gradually growing base of followers back home. These connections rely on social relationships developed prior to the migration decision, in which trust is a defining component (Waldinger 1997, Marques 2005, Heisler 1999). However, it is naïve to ignore the existence of more formal, professional networks, often taking a less altruistic form, as with syndicates.

    In this study, I would like to divide networks broadly into two categories: formal or institutional, and informal or non-institutional. Institutional networks point to the fact that once international migration is initiated and becomes a trend, private and voluntary organizations develop to support and sustain migrants (Massey et al. 1996). They provide support in a number of forms, including legal and extralegal, such as visa services, transport, labour contracting, and housing, among others, which constitute a form of social capital as they become institutionalized (Massey, Arango and Hugo 1996, Waldinger 1997). Institutional networks provide potential migrants with information, in particular about possible destinations, and sometimes with funds for transportation and contacts with agents and agencies. Abroad, they provide assistance in the form of initial accommodation or employment and mediate between the receiving society and the 'newcomers' (Dannecker 2003).

Network analysis occupies a substantial space in the current study. The significance of networks in the age of global information technology greatly increase the likelihood of migration (Nasra and Menon 1999) as they lower information costs, and, in many cases, the risks involved in the migration process. Networks are also believed to be helpful in the social integration of the newly arrived migrants into the host society, as well as hastening their entry into the job market and supporting them in their search for accommodation. Krieger (2004) observes that second generation migrants are linked with an existing network and many who rely on it for migration are less likely to incur high costs in financing migration than first generation migrants.

In explaining the other forms of network assistance in the migration process, Kritz et al. (1992) suggest that migrants gain information about migration processes through networks, formal and informal, which directs and facilitates their migration decisions. Thus, the costs and risks of international migration are significantly reduced and the likelihood of migration increases with the presence of a constellation of relationships.

An example to illustrate the growing dependence and importance of institutional networks is as follows: There are about one thousand manpower export agencies registered with the Bangladeshi government to facilitate migration. The role of these agencies is significant in facilitating Bangladeshi labour migrants, as the majority of the population has limited access to information (BMET 2005). These agents form the ring of institutional networks, which the prospective migrants rely upon to access information about the migration process and possible destinations and opportunities. Apart from agents, information about available jobs and living standards abroad is most efficiently transmitted through personal or informal networks, such as friends and neighbours who immigrated earlier or who have returned home post-migration.

Notably, in the current study, the network linkages that are active in promoting migration vary largely between the two sample populations, the Bangladeshi Respondents in Hong Kong (BRHKs) and the Bangladeshi Respondents in Malaysia (BRMs). Two figures illustrate the variation between the respective network linkages of the two sets of migrant workers. Migration to Hong Kong, for instance, is marked by a higher index of kinship ties, which are referred to as 'own network' by Maharaj (2004, Nasra and Menon 1999), rather than institutional assistance. The reason for this is that the volume of paperwork (such as visa endorsements, producing bank statements, invitation letters or sponsorship letters) for entry into Hong Kong is not as complex as in Malaysia.

While there is a general assumption among migration scholars that Malaysia recruits Bangladeshi workers through formal recruiting processes, a number of studies support that a huge volume of migrant workers enter Malaysia using illegal channels, mainly syndicates brokering their passage: Alongside the legal recruitment process, there is a significant surreptitious inflow of Bangladeshi job seekers masquerading as tourists or visiting professionals through porous land and sea entry points, guided by illegal agents (Abdul-Aziz 2001). It is clear, therefore, that the existence of networks spanning Bangladesh and Malaysia is one explanation for the continuous flow of Bangladeshi migrants into Malaysia, demonstrating that networks are not only a resource for migrants, but further, shape their destinations. The networks which have developed between Bangladesh, Malaysia and Hong Kong are one reason for the constant, almost institutionalized flow of Bangladeshi migrants to these destinations (Dannecker 2003).

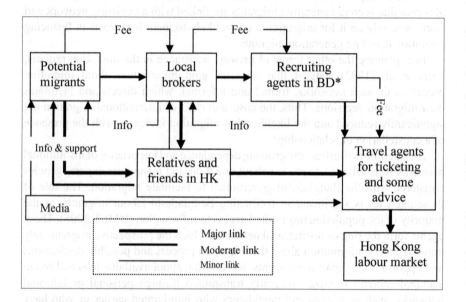

Figure 2.4 The migration networks for BRHKs
Note: * Bangladesh.
Source: Adapted from Rahman 2004.

The following figure illustrates how potential Bangladeshi migrants rely on different networks and information channels in order to migrate to Hong Kong. The implication of networks in the migration process has long been recognized, particularly in reference to the presence of relatives in the places of origin and destination who provide support in the processes of moving and settling (Lindquist 1999, Nasra and Menon 1999).

The following figure explains the networks used by the interviewed migrant workers in Malaysia (BRMs). While the network approach argues that the presence of networks reduces the cost and risk of migration, there exists a significant amount of empirical information that suggests that, in fact, network syndicates do neither, and may actually increase costs and risks in some cases. The migrant workers who relied on syndicates for their entry into Malaysia, and resorted to illegal routes through Thailand or Singapore reported facing a number of highly distressing and dangerous situations, as well as a higher cost than those who chose to migrate to Hong Kong. Therefore, the current study challenges the assumptions and arguments of network theory.

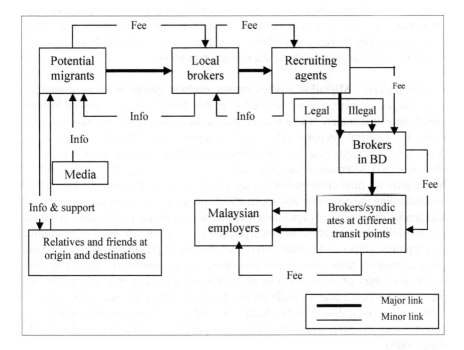

Figure 2.5 The migration networks for BRMs
Source: Adapted from Rahman 2004.

It should be noted that Figure 2.5 suffers serious drawbacks when it is applied to formally recruited migrant workers who enter Malaysia legally and directly, without traversing other countries. Therefore, the current study does not mean to generalize the entire situation, but simply to illustrate the complications of illegal migration. Figure 2.5 grapples with the major issue, previously disregarded in discussions of migration to Malaysia, namely the notable disparity between legal and illegal pathways: While the normal route to Malaysia is directly from Dhaka to Kuala Lumpur for the legal workers, illegal migrants[7] normally get to Malaysia

_____

7   ILO (1999) has clearly demarcated regular and irregular migration. A regular or lawful migrant worker is a person who (a) is granted the requisite authorizations in respect to departure from the state of nationality or habitual residence and in respect to employment

through indirect routes such as, Bangladesh–Bangkok–Songkhla (through deep forest in the back of a car or a truck); Bangladesh–Hat Yai–Sungai (on a boat, then through the forest and hills on a truck); Bangladesh–Bangkok–Yale–Sungai Golok through the forest, using a truck or van); or Bangladesh–Singapore–Malaysia (via aeroplane). Therefore, the current study will emphasize the distinction between formal and informal networks: Most of the literature on networks has confined them to informal networking (i.e., relatives, friends and neighbours either currently abroad or returned). However, networks expand as fast as migration, and where there is a market, brokers, traffickers, syndicates, and legal and illegal agencies have and will emerge to facilitate the migration process.

Gender aspects   Migration studies cannot exclude gender roles as migration is no longer a phenomenon restricted to males. Gender issues carry additional significance in a society where women play a subordinate role and have unequal rights to ancestral property (Pyle 2001, Kabeer 2004). Therefore, economic dependency is a major issue in restructuring inequalities between women and men (Kabeer 1997). Attention to the ramifications of gender on migratory processes has been a growing field in migration studies. One of the central arguments for further understanding the economic and social affects of gender on migratory processes is that gender-differentiated population mobility act as a mirror for the ways in which gendered divisions of labour are incorporated into uneven processes of economic development (Piper 2005 and 2004). As discussed previously, the migration process involves three phases: pre-departure, stay in country of destination and return to country of origin. Gender discriminatory practices and attitudes in the pre-migration phase play a significant role (Kofman 2004). Women's employment opportunities, educational levels, healthcare and access to services in their home communities are often less well advanced or provided for men's (Skeldon 2003, Piper 2005).

While females are increasingly becoming part of the international migration labour force, the majority of women in many countries are still confined within

---

in another state where such authorizations are required; and (b) who complies with the procedural and substantive conditions to which the departure and employment in another state are subject. An irregular or unlawful migrant worker is a person who (a) has not been granted authorization by the state on whose territory he or she is presently residing, or that which is required by law in respect to entry, stay or employment; or (b) who has failed to comply with the conditions to which his or her entry, stay or employment is subject (ILO 1999). However, legal and illegal migration involves another set of connotations, namely criminalization and human rights. In the words of a rapporteur at the International Conference on Migration and Crime held in Italy in 1996, the term illegal migrant (or immigrant) implies a status of criminality *ipso facto* before any judicial determination of status. So the term ought to be avoided … (Muller 1996). However, the controversy is that, to the receiving society, migrants who stay without legal documents are not simply irregular migrants; their offenses *are* often regarded as criminal and this is why they are often detained, fined and deported.

the house, often due to religious restrictions and beliefs about the preservation of women's dignity (Siddiqui 2001). I argue that, despite the debates on the representation of women, from Mexico in 1975 to Copenhagen in 1980, to Nairobi in 1985 and then to Beijing in 1995, demanding equality and the elimination of prejudice, equality remains a point of rhetoric in most of the Asian countries, including Bangladesh (Badawi 1989, Atkinson 1990). Women's primary responsibility is confined to the domestic sphere (Robinson 1983) and from an early stage of life, girls are trained to take up housework to assist their mothers and care for their siblings. Although the status situation of women had started to improve by the late nineteenth century in the labour market context in Bangladesh, typical manifestations of gender inequality are still evident, including lower participation in the labour force, concentration of female workers in low-productivity occupations (often in the agricultural and informal sectors), and lower level of wage earnings in similar occupations.

The rigidity of the gendered private-public divide is one of the most significant factors affecting the non-participation of women in decision-making and governance (Nussbaum 2003). Whilst the gradual incorporation of women into the mainstream development process during the last century has pushed the role of women from the private sphere into the public, the question remains whether this process has effectively reduced the repression of women or merely opened up new avenues for exploitation. While domestic violence against women has been a feature of both rich and poor nations, women's increased participation in spheres outside the family has endangered her external security (Tambiah 2002). Although the private domain is traditionally associated with women, it is a complex one, in which a male (assumed altruistic) head of household is responsible for the welfare and safety of all members. Women's bargaining power at the household level is restricted, typically due to lack of access to and control over resources and information, as well as low autonomy in decision making, low self esteem, low skills and education, restricted physical mobility and generally less social power as compared to men (Panda 2006). Therefore, many of the assumptions of household strategy theory are contradicted by the situation of women and their limited participation in decision-making processes at the household level.

In addition to restrictions within the private sphere, women have commonly been hindered at the public level, unable to negotiate for themselves in order to enter the field of labour migration. It is imperative to investigate how women move differently from men and how their modes of entry tend to be different, as this impacts upon their place within the labour market and their access to social services. In both North America and Western Europe where 'family reunification' is an important mode of entry, migrant women often enter as wives and dependants of men who sponsor their admission, and therefore are usually less likely than men to enter on economic or humanitarian grounds. In addition, although many migrant women (regardless of their mode of entry) do engage in paid work, they still face a gender-stratified labour market and frequently find themselves employed in the lower or even informal sectors (Piper 2005).

The issue of female migration has not gained prominence in migration literature and migration theories have not adequately addressed gender issues (Green 1995, Oishi, 2002, despite the fact that their numbers have been on the rise and trafficking has notoriously been on the increase (Gazi et al. 2001). Trafficking, often closely related to prostitution, can take various forms and involves moral, public order, labour, migration and human rights issues, as well as concerns over the proliferation of organized criminal activity (Demleitner 2001). Since the mid-1980s there has been a small, but growing focus on female migrants who undertake migration independently (Anthias 2000), not as the dependants of their father or husbands. Today they constitute 47.5 percent of all international migrants and there is increasing evidence of females migrating as principal wage earners (IOM 2004). Migration streams draw from all social groups, i.e., women and men, married and single, investors, middle class with degrees and labourers, moving within the third world and from the third world to the first (Salaff 1997, Ong, Chan and Cheu 1995).

I argue that the representation of Bangladeshi female migrants in the international labour market is the lowest among all of the countries in Asia although the total share of women in the economically active population is 39 percent. The human development gap is aggravated by substantial gender disparities with a difference between HDI rank and GDI rank of -5 (Neft and Levine 1997, Ofreneo 2000). Women have been affected by financial and social crises to varying degrees, as well as restrictions due to religion. Ullah and Routray (2003) observed that women in Bangladesh suffer the brunt of extreme poverty; however, migration propensity amongst poverty-prone women is the lowest in the country. This leads to an interesting question as to whether poverty-driven migration theory is annulled by gender in Bangladesh.

Therefore, the predominantly male migration flow from Bangladesh has left women in another problematic situation. As indicated by Afsar (2002), the workload (household work and childcare) of left-behind women increases when male members of the family migrate. Women in male migrants' households have lower labour force participation rates than those in female migrants' households. Women left behind often experience greater stress as they are over-burdened with child-care and household responsibilities, and are often concerned about extramarital affairs.

Many migrants suffer from poor health while they travel or work abroad, often due to a lack of access to medical facilities, being overworked, and poor working and living conditions. In addition, being away from one's family, friends and home can be distressing for migrants. The removal of members of the household can also have a significant impact on those who do not migrate (Zeitlyn 2006). The involvement of women in decision-making varies considerably. Women in joint families generally have little contribution in terms of generating out-of-home economic benefits for the household and therefore are usually excluded from economic decision-making. However, in nuclear families the wife of the migrant generally has greater influence over decision-making and a greater degree of freedom of choice. Economic theorizing of the household is couched in terms of decision-making but there is a divergence in the conceptualization of intra-household relations between those who deny that conflict is a factor in intra-household relations and those who allow for it (Kabeer 1997). The powerful place held by the male breadwinner in development studies has meant that the women's breadwinning role has often been neglected (Kabeer 2003).

Better living conditions in concordance with a new economic status and the changing responsibility of women in the absence of their male counterparts, as well as the disruption of marital unions have influenced the fertility behaviour of women in migrant households. In the case of wives, as the acting head of the household, the truancy of their partners results in further augmented responsibility. Labour is a valuable asset for the entire household and the importance of economically active household members is particularly crucial for financial sustainability. Families with more earning members have greater livelihood security. A trend has also been observed of young men marring before they migrate for work, to maintain a labour balance in the household. Researcher would say that a shift from 'working son' to 'working daughter' takes place in terms of labour in the family.

Further, while synthesizing the dominant migration theories, it was found that gender, while of importance, has barely been addressed. This brings to the fore the 'myth' that most migrants are men and women come only as their dependants (Green 1995, Oishi 2002). This prevailing view is predicted to decline as one study says, 'Female labour migration from Southeast Asia has become a migration stream of global significance and may eventually outstrip the significance of male labour migration from the region' (Wee and Sim 2003). Also, the trafficking of women, which intersects with migration issues, has notoriously been increasing (Gazi et al. 2001, Simic 2004).

Bangladesh, a labour intensive economy, has a weak absorptive capacity for its male labour force, let alone their female counterparts who constitute nearly half of the total population. Social and religions sanctions prevent them from going beyond their residential boundaries, albeit in recent years this practice is diminishing due to NGO intervention which includes, among other services, economic opportunities to women. However, the conventional idea of marrying off daughters to respectable families in society certainly does not allow daughters to migrate alone to other countries, with only a few exceptions. In addition, Bangladeshi women are generally not able to arrange overseas jobs by themselves due to language barriers, reluctance to be exposed and conservative beliefs. This helps to explain why samples in many migration studies are often skewed towards male respondents. The current study is no exception. In neo-classical models, gender-differentiating factors, such as the availability of marriage partners, are added in order to explain the 'extra' influences acting on female migrant populations. Usually, women are either less likely to participate in immigration because of their lack of qualifications for the labour market, because they migrate as wives and dependants of male migrants and thus are categorized as a 'special group' (Grieco and Boyd 2003).

Assimilationist and adaptation approaches: After arrival at the destination, migrants face many challenges such as changing dietary habits, climate, language, employee-employer relations etc. In such circumstances, issues of assimilation and adaptation to the new society merit understanding. Generally, assimilation is the policy of incorporating migrants into a society through a one-sided process of adaptation. Immigrants, for example, are expected to give up their distinctive linguistic, cultural or social characteristics and become indistinguishable from the majority population. In quantifying assimilationist characteristics, MacPherson and Gushulak (2004) delineated basic variables such as the volume of the impact of migration on the social and environmental capacity to accommodate an increased population load without negatively effecting the local population; as well as the degree of strangeness in cultural expression, language and religious beliefs and practices. The trajectories of migrant adaptation as envisaged by the canonical concepts of assimilation theory hold only in certain cases, largely because it espouses a container-concept of space, as though the adaptation of migrants within nation-states is a process not significantly influenced by border-crossing transactions (Faist 2000). In addition, assimilationist theory is not applicable to transient and temporary migrants.

While questions of adaptation initiate prior to migration, this theory relates best to post-migration perspectives as, practically speaking, adaptation occurs only after migration takes place. The rationalizing argument regarding migration experiences revolves around the concept of assimilation,[8] which may be cultural, political, spatial or social, and which determines how migrant workers adjust to and rationalize their place in the receiving society. Although, the migrants are often welcomed as workers and temporary sojourns, they are not so as long-term residents or settlers, as individuals, families or communities (Castles 2003). Castles (1999) provided the grounds for this fact with the following statement:

> Permanent settlements are seen as threatening to the receiving country for economic reasons such as pressure on wages and condition, social reasons such as demand on social services and the emergence of an underclass, cultural reasons such as challenges to national culture and identity, and political reasons such as fear of public disorder and effects on political institutions or foreign policy. (Castles 1999)

Assimilationist theorists (such as Castles, Vertovec, Jackson and Chiswick) address issues of rationalization because it has been found that an individual's

---

8   The idea of social exclusion model contrasts with the assimilationist idea: A situation in which immigrants are incorporated into certain areas of society but denied access to others, such as welfare systems, citizenship, and political participation are characterized by the differential exclusion model (Castles 1999, in Vertovec 1999).

level of assimilation often determines the length of residence in a destination country. Retention i.e., how long a migrant stays, is determined by – among other factors – how positively the migrant justifies or rationalizes his or her migration. Individual assimilation and adaptation is required once migrants settle in the new destination as this helps them acculturate to the host country, which ultimately aids them in obtaining employment more easily.

Therefore, according to this theory, it is important for Bangladeshi workers, as with other migrant populations, to adopt adaptation and assimilationist approaches to cope with their new environment. This is because, most often, migrants are structurally pulled into the secondary labour markets in industrialized countries, which are characterized by low wages, less preferable working conditions, and lack of job security. Host countries often consider migrants as exclusions or as transients. In the case of illegal migrants, especially, assimilation and adaptation strategies are more important, as derogatory attitudes held among the host country citizens are common. Therefore, their chances of remaining in the destination country is significantly increased by either consciously assimilating or by disguising themselves and hiding from authorities, depending on the circumstances. There is evidence that crackdowns performed to seize illegals often simply causes them to flee to the jungles to escape arrest and repatriation; although not a strategy of assimilation, certainly a form of adaptation in order to prolong their stay. Migrant workers commonly develop terminologies as an adaptation strategy to communicate among their compatriots in order to get job offers more easily and to avoid police arrest. Studies reveal that Bangladeshi migrants tend to adopt four main strategies to remain in Asian destination countries: hiding, bribing police, eliciting help from employers, and voluntary or forced confinement (Waldinger 1997).

As a part of cultural assimilation, migrant workers learn local languages, which allows greater integration into the receiving society. Due to the motivation to assimilate, learning English is not a priority for most migrant workers, instead migrant workers in Hong Kong would learn Cantonese to more easily communicate with local employers, and those in Malaysia would learn Malaysian Bahasa. In previous studies, it has been indicated that many migrant workers may try to establish romantic relationships with local women as a form of adaptation and assimilation, and may resort to marriage in order to remain in the destination country (Abdul-Aziz 2001). Ahmed's (1998) finding is quite telling in that the current suspension of Bangladeshi labour migration in Malaysia is correlated with an increasing incidence of intermarriages between Bangladeshi migrants and Malay women.[9] This is despite the fact that Malaysian regulations prohibit marriages between migrants and citizens, as is the case in many other countries (Ahmed 1998).

---

9 Kritz et al. 1992, Massey et al. 1996, Oishi 2002, Ghatak 1995, Berninghause and Seifert-Vogt 1992, Chiswick 2000, Piore 1979, Jackson 1986, Castles 2003, Castles 1999, Ahmed 2000, Ahmed 2005, Sziarto 2002.

The research methods outline

While the previous section provided the conceptualization for the current research, this section describes the operationalization process, encompassing the methods and techniques applied in the research from planning, to designing the questionnaire, selecting the field and sample, and collecting and analyzing the data. The data collection aimed at gathering information on the rationalization processes of Bangladeshi migrant workers regarding their migration and post-migration experiences.

This research is primarily based on qualitative data, supplemented by relevant quantitative data. In regards to the data sources, the study used both primary and secondary sources of data. Interviews with migrant workers were conducted using both open- and closed-ended questionnaires, and a checklist was used to collect further qualitative data. Before administering the survey, a pre-test was conducted to avoid redundancy and to add any necessary points which were previously not included in the questionnaire.

The approach used in the current study was formulated after assessing the scope and the magnitude of the research problem. I opted for an approach that included a number of levels of observation and analysis while keeping the objectives of the study in mind. Data were collected between June 2004 and October 2005. The selection of study sites was performed with maps and further rationale for selecting specific locations is discussed, followed by a detailed research design. The sampling design has also been described, followed by data collection procedures, the sources of data, and data analysis techniques.

This section describes the rationale for the selection of specific research sites. Since the research concerns a widely distributed population in three geographical settings (Hong Kong, Malaysia and Bangladesh), I adopted selective methods of data collection. Hong Kong and Malaysia have been purposively selected as study sites: Although the Bangladeshi migrant workers are very significant population in the workforce in Malaysia, very limited research has been done on them. The Bangladeshi migrant workers in Hong Kong are fewer in number, as compared to those in Malaysia, but evidence of their presence in Hong Kong goes back to the mid-1970s (BMET 2006). Until now, this population also has not received significant research attention. The specific locations of the study sites are shown on the regional maps below.

Hong Kong   Before conducting interviews, on-the-ground research was done to ascertain the magnitude of the flow of migration, the main locations of the respondents and the types of work available to migrant workers in Hong Kong. This preliminary research in Hong Kong identified four areas in which most Bangladeshi migrant workers assemble to seek work offers. Particular spots are known for specific types of work availability.

Table 2.1 Study sample sites and types of work available in Hong Kong

Sample sites	Types of work available
Sham Shui Po	Loading and unloading cargo from trucks, lorries and vans for wholesale shops and at constructions sites; factory work; house cleaning and other domestic chores
Tsim Sha Tsui	Sales agents for watch and tailoring shops or hotels; pimps
Central	Sales agents for watch and tailoring shops or hotels
Wanchai	Sales agents for watch and tailoring shops or hotels; pimps

Source: Author's field data, 2004–2006.

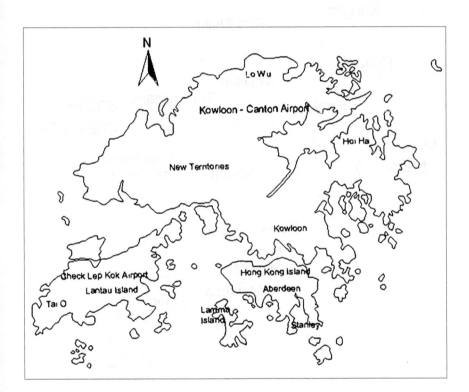

Map 2.1 Map of Hong Kong

Sham Shui Po (SSP) is geared toward a local market, and most Bangladshi migrant workers wait on street corners in the area to seek work as day labourers. Tsim Sha Tsui (TST) is more geared toward sales, and many Bangladeshi migrant workers seek work as sales agents for various hotels, tailoring and watch shops. They wait on streets corners to catch clients or employers who hire them on an hourly basis.

A few of them work as pimps for local sex workers. Central, an area located in the Central and Western district of Hong Kong Island, is the busiest district in Hong Kong. Here Bangladeshi workers are found most commonly at the ferry piers seeking work. Wanchai is mainly a business and entertainment district, where bars, dance halls, nightclubs and karaoke clubs coexist with modern office plazas, and a wide variety of restaurants, providing a number of work opportunities.

Map 2.2 Map of Malaysia

Malaysia Uncertainty about the number of Bangladeshi labour migrants in Malaysia is evident across sources: Newspapers carry estimates of illegal migrant workers based on only rough calculations, and even the few extant research studies can make no accurate estimate. According to newspapers and Malaysian police reports, more than 40,000 Bangladeshi workers were residing in Malaysia illegally at the time of this study (*Daily Jugantor* 2004). In the late 1990s, Malaysia almost closed their labour market to Bangladeshis even though a large number of workers had legal contracts that had not yet ended. The lack of precise figures for the population of Bangladeshi migrant workers in Malaysia has made it difficult to ascertain the sample frame. However, respondents were selected purposively from four major areas in Malaysia (Map 2.2). Migrants were known to concentrate in these locations after the Malaysian authorities conducted a crackdown on illegal workers in the country. Kuala Lumpur[10] is an initial destination for many migrant workers as this was the primary location of employment in the mid-1980s. Eventually, after their contracts expired, migrants became illegal and spread out to the peripheries of the country, including Shah Alam, Selangor's state capital and a modern township near Kuala Lumpur, Petaling Jaya, and five other major townships including Kelang, and Putra Jaya (the new location of Malaysia's central government). During the construction of Putra Jaya, many Bangladeshi labourers were employed, and most of their work permits expired as soon as the construction work finished. Many remained illegally and worked in construction sites and factories nearby. Johor Bahru (Map 2.2) is connected to Singapore by road and rail via a causeway. Many Bangladeshi migrants work at construction sites in Johor Bahru. They report that they chose this area because police surveillance is loosely maintained.

Bangladesh   Thirteen return migrants have been included as case studies. These have been used to triangulate the information obtained through the questionnaire surveys, and have added value and reliability to the survey results. The respondents were selected using a snowball method. Brief geographical information on the study sites is given below for understanding of the regionalization of migration.[11]

---

10   The following spots are included in Kuala Lumpur area: *Kajang* a town in the state of Selangor, located about 20 km to the south of Kuala Lumpur; Seri Kembangan and Seri Serdang, also in Selangor; and *Kelang* on the Kelang River, one of the largest cities in the country (see Map 2.2). Many small industries (shoe, paper, printing press etc.) are located in these areas. These industries require cheap labour which draws migrants. In addition, Bangladeshi labour migrants opt to stay in these areas because of cheap accommodation, less police harassment and relatively better opportunities to find work.

11   Likelihood of migration often depends on the geographical uniqueness of origin and destination locations. Migration choice is often facilitated by a constellation of relationships and links between places of origin and destination through various networks and this constellation itself creates a particular form of geographical link, cited in the literature as 'regionalization' (Neumann 1992, Chitose 2001, BBS 2004).

Map 2.3 Map of Bangladesh

Dhaka, the capital city of Bangladesh, has approximately 12.7 million economically active people in the city, with almost half of its population unemployed or underemployed (BBS 2004). Barisal, a river port (Map 2.3) has a population of 3.8 million economically active people, most of whom are unemployed. Comilla district is only 114 km from the capital and just a day's trip by road on the way to Chittagong (Map 2.3). Noakhali district (part of the Chittagong Division) is bounded by Comilla district on the north and was seriously affected numerous times by natural disasters such as high tidal bore, tornado, flood, cyclone, etc. Chittagong has 7.4 million economically active populations and is the second largest city in Bangladesh, and a busy international seaport. According to migration history and trends, both internal and international, the above districts are known to have high migratory trends among both blue-collar workers and professionals (BBS 2002, Mujeri and Khandker 2002).

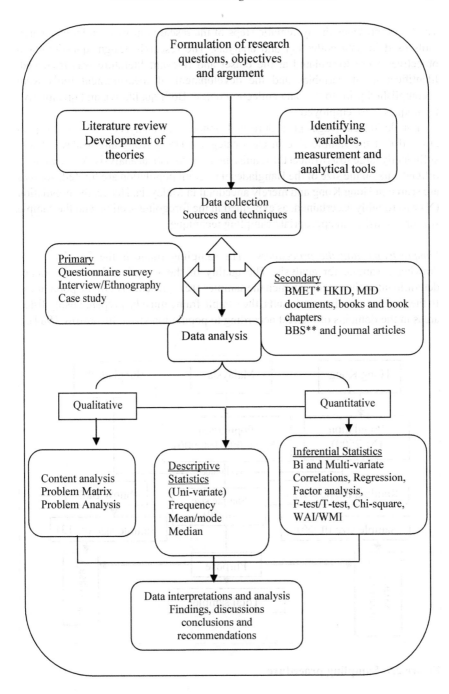

Figure 2.6 Research design framework
Note: * Bureau of Manpower, Employment and Training; ** Bangladesh.

The figure provides the systematic steps of the research process and the research tools used in data collection and analysis. The research design, questions and objectives were formulated at the outset and relevant literature was reviewed. Identification of variables and the development of measurement tools were accomplished prior to the data collection phase. Both qualitative and quantitative techniques were employed.

Population coverage, procedure and domains of analysis: The sample plan ensured that the sample size in each category (Hong Kong and Malaysia) was sufficiently large so as to meet the requirements of statistical analysis. As mentioned, official data on the size of the Bangladeshi migrant population are limited: sources are scarce in Hong Kong and merely anecdotal in Malaysia. Hence, the population (N) was roughly ascertained, as explained in the foregoing section, and the sample was drawn using an appropriate sampling technique.

Sample frame and the procedures The sampling frame is the foundation for drawing a sample for analysis. The quality of the sampling frame is a major determinant of the extent to which the samples are representative of the population. In the absence of a list of migrants, the sample frame must be prepared using listed areas in the domains of interest according to population size (Bilsborrow 1984).

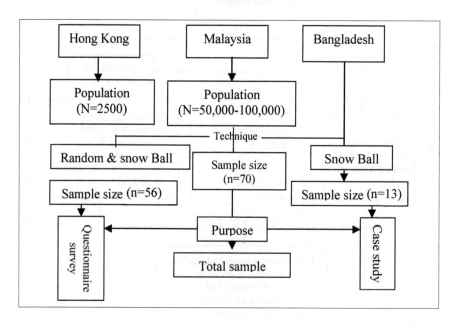

Figure 2.7 Sampling procedure

Estimates vary across sources   There are approximately 5,000 and 50,000–100,000 migrants in Hong Kong and Malaysia respectively, with an average minimum of one year of stay (Siddiqui 2003, *Daily Inqilab* 2004). According to political party leaders,[12] more than 5,000 Bangladeshis are working in Hong Kong. Around one quarter of them have settled in Hong Kong as business persons while another quarter stay as legal temporary (contractual) workers. This indicates that the rest (approximately 2,500) are migrant workers. This was crosschecked with the information provided by the Bangladeshi community in Hong Kong and by the labourers themselves, which yielded estimates of 2,000–4,000 workers in Hong Kong at the time of the study. Newspaper sources gave similar estimates (*Daily Inqilab* 2004). It should be noted, that the sample populations of Bangladeshi migrants are primarily male as cultural constraints and government policies have discouraged female migration. Therefore the study sample is also predominantly male.

Sampling technique   A number of conditions are required to determine the appropriate sample size in a study. The conditions are: the level of precision, the level of confidence or risk, and the degree of variability in the attributes being measured (Miaoulis and Michener 1976). The level of precision – sometimes called sampling error – is the range in which the true value of the population is estimated. This range is often expressed as percentage points (e.g., ±5 percent) (Israel 1992). The confidence level or risk level is based on ideas encompassed by the central limit theorem.[13] In a normal distribution, approximately 95 percent of the sample values are within two standard deviations of the true population value (e.g., mean). In other words, if a 95 percent confidence level is selected, 95 out of 100 sampled cases will illustrate the true population value within the range of precision (Israel 1992, Smith 1983). However, there is always a chance that the sample obtained does not represent the true population value. Degree of variability refers to the distribution of attributes in the population. The more heterogeneous a population is, the larger the sample size required to obtain a given level of precision. The less variable (i.e., more homogeneous) a population is, the smaller the sample size needed (Apap 2000, Cochran 1963, Miah 1999, Sudman 1976, Israel 1992).

The following equation calculates the sample size:

$$n = \frac{N}{1 + N(e)^2}$$

n = Sample size; N = Sample population; e = Level of precision

---

12   Leaders of political parties in Bangladesh and Hong Kong who claim to have accurate statistics on the total Bangladeshi population in Hong Kong.

13   The central limit theorem states that, given a distribution with a mean m and variance s^2, the sampling distribution of the mean approaches a normal distribution as N, the sample size, increases. The central limit theorem explains why many distributions tend to be close to the normal distribution.

Out of the three female respondents, one was from Tsim Sha Tsui area, while the other two were interviewed in Bangladesh using the snow-ball method. One respondent is Hindu and the other two are Muslims. They were interviewed on the condition that their identities would be protected.

Table 2.2 Distribution of samples

Study area	Sample size	Selection procedure	Data collection technique
Hong Kong	*56*	Snow ball and random	
Sham Shui Po	27		Questionnaire survey, in-depth interview, ethnography, and observation
Tsim Sha Tsui	13		
Central ferry	11		
Wanchai	5		
Malaysia	*70*	Snow ball and random	
Kajang	6		
Kelang	4		
Kuala Lumpur	10		
Ampang	6		Questionnaire survey, in-depth interview and observation
Seri Kambangan	18		
Seri Serdang	6		
Shah Alam	4		
Putra Jaya	5		
Johor Bahru	11		
Bangladesh	*13*	Snow ball	
Dhaka	4		
Barisal	3		
Noakhali	2		Case study, participant and observation
Chittagong	2		
Comilla	2		

The analysis

Questionnaire design A questionnaire occupies a significant space in any research that intends to grapple with quantitative aspects. In designing the questionnaire, issues such as efficacy in data entry, measurement, and quantification, data gathering, and ease of comprehension for interviewees were considered. Some principles were followed in order to construct the questionnaire such as: relevance of the questions to the research question and objectives; avoidance of ambiguity;

avoidance of double-barrelled[14] and leading questions, causing respondents to choose one response over another due to wording; and the avoidance of questions that were beyond the respondents' capabilities to answer, and which might frustrate the respondents resulting in poor quality responses.

Variables developed and measured   No single study can explore fully the varied dynamics of a phenomenon, hence this study has been limited to a focus on the rationalization of migration decisions, and the variables were developed carefully to this end. Migration decisions and rationalization are the dependant variables, which are affected by a number of independent variables, subsequently explained by appropriate justification for inclusion.

Sources of data (primary sources)   In order to obtain primary information I conducted structured face-to-face, in-depth interviews with the respondents using questionnaires.

Ethnography   Ethnography is believed to be a strong tool for collecting qualitative information as it involves the observation of the research participants in their common, day-to-day settings. Although formal ethnographic studies normally take weeks or even months in order for a researcher to become immersed in the setting, I was able to spend only a few weeks total with the respondents, but was still able to obtain a solid grasp of their daily activities. In Hong Kong between August and December 2004, I spent eight days at Sham Shui Po, seven days at Tsim Sha Tsui, five days at Wanchai and four days at Central with the migrant workers in order to observe their daily life. Later in 2005, I was able to spend an additional 15 days with these migrant workers. In Malaysia, I did not have an opportunity to spend time with the migrant workers in their locations of job-searching. I was, however, able to spend three days at a construction work site with the migrant workers and five days at their living places in Seri Serdang and Seri Kembangan.

Case study   I conducted the case studies using a checklist which covered: income (pre- and post-migration), occupational dynamics, vulnerabilities, asset building, and savings. The focus of the case studies was on the objectives and impact on well-being of migration.

The field research commenced in June 2004 in Hong Kong where I made efforts to gauge the extent of migration from the available statistics. I made a narrative review of 'grey' literature on various aspects of migration. Secondary sources included annual reports and documents from the Bangladesh Bureau of Manpower,

---

14   A double-barrelled question consists of two or more questions joined together which generate confusion.

Employment and Training, BBS (BMET),[15] HKID, CSD[16] and MID, SD.[17] To develop theories, contemporary literature, journal articles and selected books were extensively reviewed. Selecting appropriate data analysis techniques was a critical stage of the research process because the analytical technique applied is determined by the type, quantity and quality of data. This study applied both qualitative and quantitative analytic techniques. Qualitative data were documented, expanded and transcribed. Interviews were recorded and transcribed verbatim. Descriptive codes corresponding to the items in the migration framework identified by informants were created and assigned to the text for analysis. Effectively, the analytical process started during data collection as sequential or interim analysis has been found to be advantageous for refining questions and pursuing emerging avenues of enquiry (Pope, Ziebland and Mays 2000). Unlike quantitative analysis, which has established norms and practices for data collection and analysis, qualitative analysis is diverse in practice and offers freedom to the researcher as to the most effective methods of collection and analysis in a given situation. Given the scarcity of prior research, information and statistics, qualitative data collection appeared to be the most useful option.

Qualitative   Qualitative interviews were applied as the primary research method due to their inherent strength, depth and ability to illuminate the complexity of a phenomenon. The open structure of qualitative interviewing allows unexpected issues to emerge which relate to the aims of the study and allow insights into how migrants make decisions. For example, some interviewees may be reluctant to speak from their personal experiences and instead may prefer to discuss topics in more general terms.

Quantitative (descriptive statistics)   Descriptive statistics were used to provide an illustration of the magnitude of a particular issue. Demographic as well as socio-economic information were collected and analyzed using descriptive statistics. Throughout the book, cross tabulation, bar and pie charts and graphs have been used for data presentation to show the magnitude of the phenomenon according to the objectives of the current research. In order to describe data, mean, mode, median, percentages, standard deviations, and frequencies have been widely applied to further delve into the dimensions of rationalizing migration.

Inferential statistics   While descriptive statistics may be too shallow to infer from, inferential statistics offer a deeper explanation of a phenomenon. This research applies Chi-square (χ^2) tests and T-tests to compare a range of indicators (between the BRHK and the BRM samples; and the pre- and post-migration periods). T-tests

---

15   Bangladesh Bureau of Statistics, Bureau of Manpower, Employment and Training, Bangladesh.

16   Hong Kong Immigration Department, Census and Statistics Department, Hong Kong.

17   Malaysia Immigration Department, Statistics Department, Malaysia.

have also been applied to illustrate the differences in income, vulnerabilities, amount of asset holdings, etc. While chi-square (χ^2) tests were used for nominal and ordinal measurements, T-tests were used for interval and ratio measurements.

Multiple regression Generally, multiple regression is applied to determine what proportion of the variance of a continuous variable is associated or explained by other variables. This study applies multiple regression in order to analyze the determinants of motivations for decision-making to migrate. Stepwise-multiple regression was performed to ascertain the variables most closely associated with 'decision-making' as it is likely that a number of different factors play role to varying degrees. The outcome of the analysis enables the determination of relationships between dependant and independent variables and the level of significance for each variable. The model is set as below:

$$Y = a_0 + a_1{}^*f_1 + a_2{}^*f_2 + a_3{}^*f_3 + a_4{}^*f_4 + a_5{}^*f_5 + a_6{}^*f_6 + \ldots \ldots \ldots \ldots \ldots \ldots \ldots \ldots \ldots \quad \ldots \ldots$$
$$a_{21}{}^*f_{21}$$

Logistic model This section provides a multivariate analysis using a logistic regression for odds ratios. It requires a definition, for each variable, of a reference dimension which takes over the function as a comparator for the odds of the other dimensions. An odds ratio above the value of one indicates a positive influence, while an odds ratio with a value less than one indicates a negative relationship. All odds ratios are tested for their significance level. Only results with a significance level higher than 95 percent are deemed statistically significant. Logistic regression, in which the probability of migration was the dependent variable and age, education, experience and skill, marital status, and employment were the independent variables, was also been carried out.

Table 2.3 Variables predicting migration

P_1	Higher income	P_{12}	Desire to settle abroad
P_2	Search for work	P_{13}	Means to sneak into European
P_3	Pressure from family		countries
P_4	Induced by brokers	P_{14}	Means to settle a disputed piece of
P_5	Induced by relatives staying abroad		land (sale)
P_6	Joining relatives	P_{15}	Means to use savings
P_7	Information on foreign countries	P_{16}	Nothing to do in home country
P_8	Stubborn desire to work overseas	P_{17}	No education/no training
P_9	Threatened by political forces	P_{18}	Marital factor (unmarried)
P_{10}	Escape conviction	P_{19}	Marital factor (married)
P_{11}	Inadequate information	P_{20}	Loss in previous business
		P_{21}	Reluctance to work in paddy fields

Source: Author's field data, 2004–2006.

Factor analysis   Factor analysis was performed in order to discern the common predicting variables for migration. It extracts only that proportion of variance, which is due to the common factors and shared by a selected set of variables. The proportion of variance for a particular variable that is due to common factors (shared with other variables) is called a communality. The eigenvalue for a given factor reflects the variance in all the variables, which is accounted for by that factor. The ratio of eigenvalues is the ratio of explanatory importance of the factor sets with respect to the variables. Factors having a low eigenvalue contribute little to the explanation of variances and are largely ignored.

Table 2.4 Problems faced in post-migration

f_1	Bribing police	f_{18}	Locals look down on them
f_2	No long term contract, no job security	f_{19}	Very low income/low salary
f_3	Exploitation and inhuman behaviour from the employer	f_{20}	Cannot go back home due to lack of money/savings
f_4	Employers confined them	f_{21}	Irregular salary or employers hold the salary
f_5	Cut levy from salary		
f_6	Hold their passport	f_{22}	Salary never increases
f_7	Life risk at work	f_{23}	Salary lower than promised
f_8	Risk of imprisonment	f_{24}	Language barrier
f_9	Oppression by Tamils	f_{25}	No time to cook and clean up
f_{10}	Insulted, such as being called '*orang* Bangla'	f_{26}	Limited health access
		f_{27}	Banks do not give check books
f_{11}	Snatch theft by Tamils	f_{28}	Money cannot be withdrawn without employer's letter/consent
f_{12}	Taxi drivers and bus conductors charge more than for locals		
f_{13}	Deception by clients	f_{29}	Money has to be deposited at home (in Bangladesh) as no access to local banks
f_{14}	Beaten by Tamils or Malays for no reason		
f_{15}	No place to lodge their complaints against any injustice	f_{30}	Long queue for toilet/bathroom at living place
		f_{31}	Employers do not give time for prayer
f_{16}	Restricted movement	f_{32}	Employers sometimes force to work
f_{17}	Many boarders in a single room	f_{33}	Reproductive related

Source: Author's field data, 2004–2006.

Index of loss and benefits   Losses and benefits due to migration are the principal determinants of the rationalization of the migration decision. The perceived loss or benefit experienced by respondents was measured by applying the Weighted Mean Index (WMI). Therefore, degree of perceived benefit from migration was measured using the WMI. This helps to justify and rationalize their migration decision in a quantitative manner.

$$\text{WMI} = \frac{(w_i f_i + w_2 f_2 + w_3 f_3 + \dots\dots\dots\dots\dots + w_n f_n)}{(f_1 + f_2 + f_3 + \dots\dots\dots\dots\dots + f_n)} = \frac{\sum w_i f_i}{\sum f_i}$$

Where,

w_i is the assigned weight for a particular class under the benefit scale and f_i is the corresponding frequency for that class. This research considers – under the degree of benefits – a five point-scale: very high, high, neutral, low and very low. The corresponding weights are 1.0; 0.80; 0.60; 0.40; and 0.20. The problems mentioned by the respondents are listed in Table 2.4.

Scale of agreement A point scale has been used to prioritize the variables based on values assigned to a particular variable. A five point-scale was used as it allows the recording of positive as well as negative responses indicating agreement and disagreement respectively (Miah 1999). The assigned values for particular responses are as follows:

Strongly disagree	Disagree	Neutral	Agree	Strongly agree
-2.0	-1.0	0.0	+1.0	+2.0

Indices of rationalization-triggering forces for migration Indexing the triggering forces for migration categorizes the variables according to the value of the Weighted Average Index (WAI). In the case of the current study, the WAI of migration helped to rank the triggering forces for migration in order of their accumulated weight.

WAI = [1st rank (1.0) +2nd rank (0.8) +3rd rank (0.6) +4th rank (0.4) +5th rank (0.2)]/ $\sum$ ith rank

Where,

1st rank f* of 1st rank COM**
2nd rank f of 2nd rank COM
3rd rank f of 3rd rank COM
4th rank f of 4th rank COM
5th rank f of 5th rank COM

* Frequency; ** Cause of migration (note: rank indicates the priority of the factors). SPSS software has been used to analyze both qualitative and quantitative data.

Reliability issues and limitations

Reliability   A quality research study has to have a degree of reliability in terms of its data and analytical appropriateness. To ensure the quality and reliability of the current research, a number of measures were adopted. Before administering the interviews, a field test of the questionnaire was performed and data were cross-checked. I conducted a rough census and in any case where a lack of statistics was identified, I tried to validate the available government data with newspaper estimates. To ensure a systematic sample, an appropriate technique was applied to determine the sample size from the population (N). I revisited the fields frequently during the entire period of research and data collection.

Limitations   Locating respondents was the primary difficulty in drawing the sample, both in Hong Kong and in Malaysia. Migrant workers in Malaysia were very much scattered in diverse and often distant locations. When I approached them for an interview they commonly declined, at first, to share their experiences. Many of them declined altogether. I can only assume that I have missed important information by not having the opportunity to interview them. In Hong Kong, I had to interview on the streets where the migrants wait for work offer. Many times respondents would rush to employers to seek work half way through the interview. Many were not comfortable to speak freely on the streets as they were anxious about police surveillance. In many cases, I was not able to complete an interview in a single day. All of these factors compromised the validity of the data, but all are common issues in field research.

Circumstances like undocumented and clandestine migration, trafficking in persons, and overstaying in host countries after a contract has expired make it complex to record the flow of migration, both in the countries of origin and destination. This resulted in data scarcity which placed a significant constraint on my research. Added to this scarcity is the dearth of female migrants which forces research documentation to be skewed to the experiences of males. Though the sample size is good enough for statistical analysis, inclusion of more respondents could have added more values to the resulting inferences.

Chapter 3
Socio-Economic and Demographic Profiles

The purpose of this chapter is to provide a descriptive summary of the socio-economic and demographic profiles of the respondents (Bangladeshi respondents in Hong Kong (BRHK) and in Malaysia (BRM) of the research the book deals with. This basic information on the respondents and their households is crucial to the interpretation of the study findings. Background profiles help to show the disparity between pre-migration milieu and post-migration experiences. Both the BRHKs and the BRMs migrated with similar expectations for economic gain. However, these expectations remain predominantly unrealized; the extent to which depending on the country of destination. The profiles comprise the socio-economic and demographic variables of the respondents themselves and of their households, including: gender, ethnicity, income, dependency ratio, ownership of home and other assets, educational qualifications, occupations (including pre-migration occupation), language, religious affiliation and skill composition (Silvey 2001 and 2000). The demographic variables include: age, family size, marital status, mobility, location of residence (districts of origin), and life cycles (fertility, mortality and migration). This chapter has two sections: the first discusses socio-economic variables, and the second deals with the demographic profiles of the respondents.

This section includes the variables[1] that describe the respondents' socio-economic profile. The gender distribution of the BRHKs shows a highly skewed pattern toward males. The data show that 95 percent of the BRHKs are males, with only 5 percent females in the sample. Although women's participation in various economic sectors has increased in the recent years in Bangladesh, their migration out of Bangladesh remains insignificant.[2] Female migrants represent 0.98 percent of the total migration from Bangladesh (as of 2009) (BMET 2010). Therefore, the slanted gender distribution in the sample is representative of the greater trend.

---

1   The importance of these variables is evident when we consider the human capital endowment in the study of migration, which posits that individuals derive economic benefits from investment in people, which refers mainly, although not exclusively, to education, skill, health, and work experience (Adams 1993, Bratsberg and Terrell 2002, Friedberg 2000, Casarico and Carlo 2003, Zeng and Xie 2004, Guerrero and Bolay 2005).

2   This may be viewed in comparison to Filipino women, who are spread over 181 countries around the world with remarkable representation in positions other than dependant migrants (KAKAMNP 1998). These women have been able to search out opportunities for themselves in the global labour markets. In Bangladesh, females constitute around half of the total population, however the ratio between male and female migrants is as wide as 100:0.98 (Khanum 2005, GoB 2005).

Religious conservatism and restricted mobility restricted Bangladeshi women from obtaining overseas jobs.

With regards to the religious affiliation of the respondents in Hong Kong, 91 percent are Muslims and 9 percent are Hindus. As with the gender distribution, the skewed division by religious affiliation in the sample reflects the actual make-up of the Bangladeshi migrant population in Hong Kong. In terms of the education of the respondents in Hong Kong, this section makes the following distinctions for educational attainment: those who finished primary education (five years of schooling), those who finished junior high school (six–eight years of schooling), secondary school (10 years of schooling), college (12 years of schooling), and university education (16 years of schooling).[3] The data show that almost half (47 percent) of the BRHKs reported having primary level education with seven percent saying that they had secondary level education; and 43 percent reported having college level education. Five percent of them had university level education and none were illiterate.

In parallel to, and for the same reasons as, the BRHK sample, 95 percent of the BRM sample are males, and 5 percent are females. Only 1 percent of the BRMs is Hindu; the overwhelming majority are Muslims, also mirroring a similar distribution to the BRHKs. With regards to the educational attainment of the BRMs, 27 percent reported having primary level education, 10 percent reported having completed junior high level education, and around 38 percent had secondary level education. The highest percentage of the BRMs had secondary level education (39 percent) followed by primary level. However, none of the BRMs was illiterate. There were also no BRMs with university education. There is no significance difference in the attainment of education between the BRHKs and the BRMs.

This section presents the economic profiles of respondents which can be assessed by examining personal income, occupational status, labour force within households, dependency ratio, household assets and individual skills (Lucas 2001). Income, economic resources and social status tend to be interrelated and economically motivated people act according to the perceived highest economic outcome (Olsson 1965, Ainsaar 2004). Here I also include the pre-migration occupational dynamics of the respondents. This shows that the occupational change associated with the transition from the origin to the destination country, not independently. This section examines three categories[4] of pre-migration occupations: employed,

---

3 From a human capital perspective, it is assumed that higher levels of education offer increased income returns for specific segments of the labour market. It is also argued that higher levels of education provide a greater ability to collect and process information, which lowers the risk and therefore increases the propensity of migration. However, Krieger (2004) finds an insignificant or even negative coefficient between levels of education and propensity to migrate.

4 The logistic regression of the inclination to migrate supports results of different causal patterns of migration to Hong Kong and Malaysia. Overall, the model includes only three significant causal relationships. None of these are surprising and all are well argued in various concepts of migration. Lastly, unemployed people are more strongly

unemployed and student. Around half of the respondents (43 percent) were unemployed before they migrated. Those who were employed (29 percent) were engaged in five categories of activities: five percent of respondents worked as representatives of the American Life Insurance Company (ALICO), 18 percent owned petty businesses, five percent worked as cooks in Chinese restaurants, 11 percent as factory workers and technicians, and around 13 percent of respondents worked as assistants in groceries and restaurants. Five percent were still studying. Further analysis on the educational attainment and corresponding occupational dynamics is worthwhile: While about half of the respondents had no employment and the other half also had no permanent job, and were only employed for survival on temporary basis, the preceding section has shown that 31 of the 56 BRHKs had higher than secondary-level education and three were university-educated. The following section explores their income earning through their employment.

In the economic profiling of the respondents, I have highlighted previous employment dynamics. Many studies have found that the 'employment' factor correlates with migration tendency. Forces of migration, such as hopes of employment as a pull factor and unemployment as a push factor, are traditionally considered very strong. As discussed in the previous section, this study distinguishes three categories of occupation of the respondents before their migration: employed, unemployed and student. Unemployment was the most common status for most of the BRMs: Data show that a majority of the migrants (67 percent) had been unemployed before moving to Malaysia. Around 9 percent were studying before they moved and the other 25 percent were employed (19 percent owned petty business, 6 percent were factory workers). The study further shows that the mean year of involvement in previous employment activity was around two years for the BRMs. While no significant difference is observed in the occupational dynamics between the BRHKs and the BRMs, unemployment status varied distinctly: only 43 percent of the BRHKs reported being unemployed prior to migration, as compared to 67 percent of the BRMs.

Income benefits from pre-migration occupation: The level of income earned is directly related to the level of education and the occupations the migrants were engaged in. Income is derived from capital, from labour, or from both combined. Income has an additional correlation to migration propensity: Migration often takes place when the perceived income benefits obtainable from migration are expected to be higher than current income benefits. Although income determines

---

motivated to migrate than other groups in the employment status category. Uncertain employment prospects are an important push factor. Labour market theory emphasizes income differentials as the strongest influence factors on migration and employment differentials as the second strongest (Ortiz 1992). Economic concepts, focusing on search and information costs, predict that unemployed people have less constrained time budgets for preparatory search and information behaviour related to migration. This would suggest a higher propensity for migration by unemployed people compared with employed people (Krieger 2004).

the economic status of households, the level of income does not necessarily reflect the 'true financial condition' of the migrants, which is often dependent on the expenditure pattern of the respective household and their consumption behaviour. Table 3.1 presents yearly pre-migration incomes of the respondents. These data examine the relationship between respondents' intended migration and their previous occupation and income. Data show that around 68 percent of the BRHKs had yearly incomes below Tk. 5,000 (US$74) and around 8 percent had yearly incomes of between Tk. 35,000 (US$515) and 40,000 (US$588).

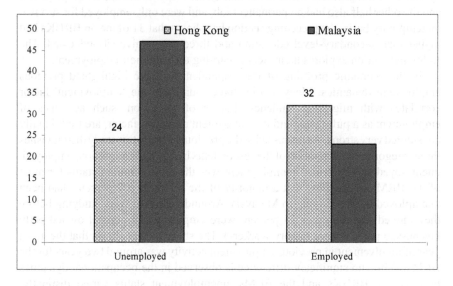

Figure 3.1 Pre-migration occupation of the respondents
Source: Author's field data, 2004–2006.

In order to further analyze the positioning of the respondents on the economic ladder in Bangladesh, a comparison was made between the official cut-off point of income and the real income benefit of the migrants. As shown in Table 3.1 the average yearly income was Tk. 9,144.74 (US$134). The cut-off point (threshold) of income for the poverty line per household is (6896*5.87)[5] Tk. 40,479.52 (US$595). This indicates a significant discrepancy, indicating that the majority of these households live far below the poverty line. According to most of the respondents, people with sufficient income to meet financial needs at home, are

_____

5   As Ullah and Routray (2003: 86) used Tk. 6,896 (US$101) per capita income as the threshold for the poverty line in Bangladesh, I am going to use this poverty index to describe the position of respondents, based on their level of pre-migration income, to provide an explanation of how their economic status relates to migration decision-making.

less likely to migrate for overseas. This is in concordance with the fact that poverty is classically seen as another of the motivating push factors for migration (Skeldon 2003, Rahman 2004).

Income is the most cogent factor in determining the echelon of the household in a social order. Data show that only 23 of the total BRM (33 percent) respondents claimed that they had a source of income prior to their migration; leaving 67 percent unemployed before their move. Data show that around 48 percent of the BRMs (the majority of respondents) fall in the lowest income category and conversely, the lowest number of BRMs (n=70) falls in the highest income category.

Table 3.1 Yearly incomes prior to migration

Yearly income (BDT)		%
Respondents in Hong Kong	**n=38**	
<5000 (US$74)	26	68.42
5001–10000 (US$74–147)	3	7.89
20001–25000 (US$294–368)	3	7.89
25000–30000 (US$368–441)	3	7.89
35001–40000 (US$515–588)	3	7.89
Mean income	9144.74	
Total	*38*	*100.0*
Respondents in Malaysia	**n=23**	
<5000	11	47.8
5001–10000	4	17.4
20001–25000	6	26.1
25000–30000	1	4.35
35001–40000	1	4.35
Total	*23*	*100.0*

Source: Author's field data, 2004–2006.

Apart from the earnings through paid work by the respondents, I have taken 'other sources' into consideration to get a complete picture of the total income earnings of the respondents before their migration. This carries importance in further analysis because income diversification or maximization to benefit poor households is becoming very common in rural areas in Bangladesh (Mujeri and Khandker 2002). Households adopt multiple income sources to reduce the risk associated with fluctuations in any single income source as it offers complementarities between activities. Income benefits from other sources[6] considered in the study are: agriculture, business and the work of non-primary household members. As shown in Table 3.2, 57 percent of the BRHKs had an annual household income from sources other than the primary earner's employment of below Tk. 20,000 (US$294). Sixteen percent of them had an annual household income from other sources of between Tk. 40,000 (US$588) and Tk. 60,000 (US$882). Four percent of respondents belong to the highest income category, due to other sources, of Tk. 140,000 to Tk. 160,000. The data show that diversified income sources contributed, in some cases significantly, to the total income package of the households.

Table 3.2 Yearly household incomes from other sources*

Income (Tk.)		%
Respondents in Hong Kong	**n=56**	
No income	2	3.6
<20000 (US$294)	32	57.1
20001–40000 (US$294–588)	9	16.1
40001–60000 (US$588–882)	3	5.4
60001–80000 (US$882–1176)	4	7.1
80001–100000 (US$1176–1471)	2	3.6
100001–120000 (US$1471–1765)	1	1.8
120001–140000 (US$1765–2059)	1	1.8
140001–160000 (US$2059–2353)	2	3.6
Mean	32133.33	
SD	37533.427	
Total	*56*	*100.00*
Respondents in Malaysia	**n=70**	
<20000	66	94.28
20001–40000	3	4.29
40001–60000	1	1.43
Mean	8058.46	
SD	8763.37	
Total	*70*	*100.0*

Note: * Agriculture, petty business, and wage earning from other family members.
Source: Author's field data, 2004–2006.

---

6   Refers to earning from other members of the households who are involved in economic activities with sources such as agriculture, business etc.

This pattern is similar, though less pronounced, among the BRM sample. Apart from the previous employment, in many cases other sources have also added extra income to the total household income earned by the BRMs. Table 3.2 shows that 94 percent of the BRMs reported having annual incomes due to additional sources of below Tk. 20,000; 4 percent had income from other sources in the category of Tk. 20,000 to Tk. 40,000; and 1 percent reported annual income from other sources in Tk. 40,000 to Tk. 60,000 category. From the data, the income sources of the BRHKs appear to be relatively more diversified than the BRMs. There is also a significant difference in the mean earnings between the two samples, with BRHK's reporting more household income from other sources. However, in both samples, even with these additional sources of income, the mean annual household income is much lower than the marginal income earning determined for the poverty line in Bangladesh (Tk. 6896*4.9 = 33790.4 (US$497)).[7]

Ownership of household assets

In the economic profile, ownership of household assets has been included because it is an important determinant of a Bangladeshi family's level of financial security (Siddiqui 1999).[8] Household asset accumulation generally tends to be under-reported in household surveys. A household's economic well-being depends on both its income earning and its asset accumulation. While income is the flow of resources to a household, asset accumulation indicates the level of security during financial vulnerability. Measuring wealth due to assets is more complex than measuring income: The market value of some assets, such as that of a house, is difficult to determine precisely. Some assets may be targeted for future consumption and are thus excluded from the analysis. Some assets, such as household durable goods, are generally not considered subject to sale. Additionally, these assets are generally difficult to value and may have an intangible value not reflected in the market. This section looks at the value of interest-earning assets, of mortgages held by sellers, and of vehicles, if any, as well as the self-reported value of owned businesses, or professions held by household members. The section also examines the household net worth, defined as the sum of assets less liabilities, such as debts secured by any asset, bank loans, and other unsecured debts.

Major assets of the respondents' households include: land holding, cattle and other valuables, e.g. agricultural equipment. This section determines the current value[9] of BRHKs household assets and separates them into two possible categories:

---

7   This is a common measure of the poverty line in Bangladesh, see Ullah and Routray 2003.

8   Both qualitative and quantitative characteristics of the households (size, quality, design, cost, and ownership) in both places of origin and destination are considered important in all migration events (Clark and Onaka 1983, White 1990, Ainsaar 2004).

9   Current value of the asset was worked out according to the following equation: $\{(CV = \sigma^{a} - \sigma^{b})$ Total previous asset – sold out asset during migration = current value-CV$\}$.

Tk. <100,000 and Tk. 100,001–200,000. Most of the respondents (95 percent) have household assets below Tk. 100,000, while the remaining 5 percent have or had assets worth Tk. 100,000–200,000. It is not possible to determine precisely the position of the household in the local society by the value of the assets owned; however, it is evident that a significant number of the respondents have or had assets of low value, indicating a low socioeconomic status in the local society.

Household asset values for the BRMs has been categorized into eight groups (see Table 3.3). The highest percentage of the BRMs (34 percent) reported having major assets cumulatively worth less than TK. 100,000 (US$1,471); the lowest in the sample. Among the eight groups, only one percent of the sample reports having the highest value of assets. Across the BRM sample, the mean worth of household assets is Tk. 251,571.43 (Table 3.3).

Table 3.3 Current value of major household assets

Value in Taka		%
Respondents in Hong Kong	n=56	
<100000 (US$1471)	53	94.6
100001–200000 (US$1471–2941)	3	5.4
Mean	42000	
Standard deviation	35556.796	
Total	*56*	*100.0*
Respondents in Malaysia	n=70	
>100000 (US$1471)	24	34.3
100001–200000 (US$1471–2941)	15	21.43
200001–300000 (US$2941–4412)	10	14.29
300001–400000 (US$4412–5882)	9	12.9
400001–500000 (US$5882–7353)	6	8.6
500001–600000 (US$7353–8824)	2	2.9
600001–700000 (US$8824–102941)	1	1.43
700001–800000 (US$102941–11765)	3	4.29
Mean	251571.43	
Standard deviation	204454.99	
Total	*70*	*100.00*

Source: Author's field data, 2004–2006.

Skill composition

This section discusses the skill composition of the respondents. Worth mentioning is the four classifications of migrant labour made by the Bureau of Manpower, Employment and Training (BMET): professional (doctors, engineers, nurses and teachers), skilled (manufacturing or garments workers, drivers, mechanics, and heavy machine operators), semi-skilled (tailors, masons, carpenters) and less skilled (domestic workers, cleaners and all other kinds of labourers). Women are more commonly employed as nurses, garment workers, manufacturing labour and domestic workers (BMET 2005).

The type of skills that migrant workers are likely to attain in order to improve their earnings position abroad often depends on the requirements of the host country's labour market, as well as on the type of skills and the skill level the migrant worker had upon arrival (Dustmann 1994). The question asked of respondents was whether they had any skills training[10] before they migrated and, if so what the training was. As shown in Figure 3.1, an overwhelming majority of the BRHKs (89 percent) had no skills training while the remaining eleven percent claimed to have some training in electrical mechanics and computer operations. However, even those who had training reported that their training was useless because the available work does not require their trained skills. Data from BMET show that of total migrant labour force in Hong Kong (from 1976–2005), 4 percent were professional, 33 percent were skilled, 16 percent were semi-skilled, and 47 percent were unskilled (BMET 2006). This means that most of the respondents are unskilled and even those having skills, there is no clear benefit since the market demand is for unskilled labour and the skills training that they have is not applicable to the jobs they are offered.

Like the case of the BRHKs, an overwhelming majority of the BRMs[11] (91 percent) did not receive any skills training before they migrated. Only six of the

---

10   The level of skill of an individual determines his/her wage rate. Unskilled labourers in Bangladesh constitute the highest composition of international migration (according to data from BMET 2005). Bangladesh has recently experienced emigration of more unskilled and semi-skilled migrants, whose wages are lower, compared with the previous wave of skilled emigrants; simultaneously, wage rates in destination countries have fallen drastically in the last decade (Siddiqui 2003). Abdul-Aziz (2001: 12) in his comparative studies (between Bangladeshi and Indonesian migrant labour in Malaysia) found that the Malaysian employers are least likely to engage Bangladeshi migrant labourers in their firms due to their relatively poor skills. He also explains that this low skill level could be attributed to the shorter duration of stay in Malaysia of the Bangladeshi workers, as compared to the Indonesian workers.

11   The primary impulsion for both the governments of Hong Kong and Malaysia to allow the importation of foreign labour has been the labour shortages experienced by these countries. Chinese mainlanders are the traditional source of labour in Hong Kong (Chan and Abdullah 1999), while for Malaysia, with diverse ethnicity (Malay, Chinese and Indians form the major compositions of Malaysian population), migrants meet the demand

BRMs claimed to have some sort of skills training. When they were asked what kind of training they received: they said it was in computer operation and driving. As with the BRHK sample, all of the trained BRMs reported that their training was useless because their skills could not be applied in the work available to them. The respondents added that their training has neither enhanced their wage rate nor did it help them in getting a better job. Abdul-Aziz (2001: 12) in comparative studies (between Bangladeshi and Indonesian migrant labour in Malaysia) found that Malaysian employers are least likely to recruit Bangladeshi migrant labourers in their firms due to their poor skill level. However, other studies show that Bangladeshi unskilled workers achieve efficiency and training on the job, therefore, this poor skill level could often be attributed to the shorter duration of stay in Malaysia of the Bangladeshi workers as compared to the Indonesian workers.

This section highlights the demographic and geo-demographic profiles of the BRHKs, based on age, household composition, marital status, mobility, location of residence (districts of origin), life cycles (fertility, mortality and migration), and habitat conditions (Andrew 2004, Malynovska 2004). The likelihood of migration often varies due to unique geographical characteristics of origin and destination, for example, people of disaster prone districts tend to have increased potential propensities to migrate (Chitose 2001). Further, the choice to migrate is facilitated by a constellation of relationships and links. This constellation is geographically situated, usually in regional contexts (see Neumann 1992). Therefore, the 'regionalization of migration' connotes that there is a regional constellation of linkages, constructing chains of migration (BBS 2009).

Demographic profiles focus on population dynamics which include race, age, income, mobility, and location as previously mentioned. The preceding sections have justified the reasons for including these variables in demographic profiles. To examine the 'ages' of the respondents as part of their demographic profile, this study categorizes five age-range groups. The data show that, in the BRHK sample, the highest number of respondents falls in the mid-age group (35–45) and around a quarter of the sample falls in the 25–30 age category. The mean age was 34.1 years – within a range of 20 to 55 – which means that respondents were relatively young when they migrated from Bangladesh. A significant decrease is observed in migration among those who are 40–55 years old. This finding resonates with the findings of other studies, such as Miyan (2003) and the IOM (2005: 191) and with the gravity laws of migration (Ravenstein 1885).

The age profile of the BRMs roughly paralleled that of the BRHKs, above. Approximately 27 percent of the respondents belong to the youngest group (25–30 years old) and the highest percentage of respondents belong to the second age group (30–35 years old). As age increases, the percentage of the respondents decreases, demonstrating a lower propensity of migration among the older respondents. The mean age of the BRMs is 33 years.

---

of labour shortages in a variety of economic sectors, because the local people are averse to taking up low esteemed work (Devanzo 1982, Chitose 2001).

The responses for marital status show that 43 percent of respondents are married and 46 percent are unmarried while a few respondents declined to report their marital status. There are fewer propensities for migration where there is strong integration into existing social networks in the sending countries compared to a low potential for social integration in the receiving countries (Krieger 2004). This section concentrates on push or retention (pull) factors in the home country. In this context, being married or cohabiting is seen as a prohibitive factor for migration, while being single is seen as an enabling factor. This is in line with socio-economic micro concepts, predicting a lower propensity to migrate where both partners work and a higher propensity for single people (Hugo 1998 and 1997, Krieger 2004). Although some other studies have shown that the likelihood of migration of unmarried men is higher than it is for married ones, this study found no noticeable difference in the migration propensity between the married and unmarried people (as will be demonstrated later in the chapter).

Two categories of responses for marital status of the BRHKs were identified: single and married.[12] Data show that 46 percent of BRHKs were married and 54 percent were unmarried. Another question was asked 'were you married when you left your country?' Data further show that during their departure to Malaysia, 29 of the total 38 who reported currently being married, were married prior to migration, the remaining nine got married post-migration, either in Malaysia to Indonesian or Malaysian women, or in Bangladesh during their stay in Malaysia, to spouses who later joined them. Three of them went to Bangladesh on holidays and got married. Two BRMs brought their wives to Malaysia and later deserted their wives after becoming engaged in romantic relationships with Indonesian girlfriends.[13] One of these wives returned home by acquiring travel cost in the form of charity from the Bangladeshi community in Malaysia after the last Mercy[14] which was declared by the Malaysian government in 2005. The other former wife remained in Malaysia after marrying another Bangladeshi man.

---

12   There are fewer propensities for migration where there exists a strong integration in existing social networks in the sending countries and a low potential for social integration in the receiving countries (Krieger 2004). This section concentrates on push or retention (pull) factors in the home country. In this context, being married or cohabiting is seen as a prohibitive factor for migration, while being single is seen as an enabling factor. This is in line with socio-economic micro concepts, predicting a lower propensity to migrate where both partners work and a higher propensity for single people (Krieger 2004).

13   These two deserted wives become helpless and later found their way of living by offering sex primarily to Bangladeshi clients. One of them married to a Bangladeshi labourer again who used to help her to manage clients. The other woman tried to manage a work but failed and later decided to engage in sex working. They were not shy to disclose their occupation as they were well-known in Sri Kambangan area in Bangladeshi community.

14   The Malaysian government offers travel passes (under a Mercy policy) to illegal migrants. They declared Mercy for the third time in March 2005.

This section demonstrates that household composition[15] is a significant element of the demographic profile. Household composition determines the 'dependency ratio (DR)'[16] which, according to many researchers, has direct correlations with the likelihood of migration (Acharya 2003). Data further present the number of income-earners[17] in respondents' households. Around 61 percent of the respondents reported living in a 6–10 member household, and 34 percent in 1–5 member households. Furthermore, the number of income-earners in a household determines the level of well-being of the household. My data show that approximately 64 percent of BRHKs reported having two earning members, while 36 percent reported having only one earning member in their household. The following offers interesting information for comparison, indicating that approximately 61 percent of BRHKs live in 6–10 member households. Therefore, despite the number of wage earners in the household, average household size for BRHKs is 5.87.

The average size of the BRM households is 6.39 (significantly higher than the countrywide average of 4.9 in 2003). Data show that 44 percent of respondents reported living in a 1–5 person household; households of 6–10 persons accounted for 49 percent of BRMs; households of 11–15 persons accounted for four percent of the sample, and households of 16–20 persons accounted for the remaining three percent. As for income earners, approximately three percent of BRMs reported having no income earners in the household, 24 percent reported having a single income earner; 57 percent reported having two income earners, 9 percent reported having three income earners and the remaining seven percent reported having four income earners. The previous section discusses why and how the dependency ratio was performed. Data confirm that the dependency ratio for BRMs is also much higher than the average countrywide ratio.

Dependency ratio (DR): In households, economic dependants are children who are too young to work, individuals who are too old or infirm, as well as those who are unemployed.[18] The number of dependants is divided by the number of

---

15   Theoretically, a migration decision is not made individually; therefore, the opinions of family members are often important. The main form of family participation in the migration decision is family members' help or at least consent, in obtaining money from different external sources to cover the migration cost (Massey et al. 1996).

16   The dependency ratio is the ratio of the economically dependent part of a population (inactive), in this case, a family, to the productive part (active).

17   The number of income earners in a family, total number of family members and the amount of family earnings are the central factors that determine the economic status of a household. They determine whether the household is better off or not in relation to other households in the reference area and when such comparison occurs the tendency to sense 'relative deprivation' emerges, which may push people to migrate in search of better opportunities and an improved lifestyle (Stark and Taylor 1991).

18   Generally, individuals under the age of 10 and over the age of 65 are considered economic dependants. This ratio becomes particularly important as it increases, i.e. there is increased strain on the productive part of the population to support the economic dependants.

wage earners in the household in order to achieve the dependency ratio (equation below). The larger the dependency ratio is in a population or household, the higher the proportion of those who are not employed, and the lower the income of the household. The lower household income is, the higher the likelihood of household members migrating from their place of origin in search of alternative, more lucrative, employment opportunities. The dependency ratio of the sending families has been worked out by using the following equation:

$$i = \frac{x}{y} \times 100$$

i = dependency ratio
x = economically active population or productive part; and
y = economically inactive population or dependant part.

Previous overseas mobility

The intensity of mobility[19] of the respondents has been analyzed using three directions of movement: current place of residence (at the time of survey; either Hong Kong or Malaysia), where they lived immediate before moving to Hong Kong or Malaysia, and place of birth. Fifteen of the BRHKs travelled and worked in other countries, such as Malaysia (n=6), Taiwan (n=3) and South Korea (n=6) before they came to Hong Kong. A few of them had visited South Korea and Taiwan only a few months before they were interviewed on a 'wait and see' approach to explore opportunities and determine whether conditions are more conducive there, particularly in terms of the availability of employment and police surveillance than in Hong Kong.

For many of the BRMs, Malaysia has not been their first or only destination. Eleven of the respondents had either visited or worked other countries prior to Malaysia. Of them, three used to work in Saudi Arabia, one in Korea, two in Taiwan and the remaining four in the United Arab Emirates. Recent imposition of strict regulations on the immigration policies and the stalemate in manpower export to Malaysia from Bangladesh, has resulted in illegal[20] migrants considering Hong Kong as a better possible choice. Therefore, some Bangladeshi migrant labourers in Malaysia initially landed in Hong Kong.

---

19   Previous overseas experience (which refers to respondents' experience before moving to Hong Kong and Malaysia) and the intensity of migration could be connected to the learning process; acquiring better knowledge or skills (Ainsaar 2004).

20   'By definition, an immigrant is illegal if s/he contravenes the law by entering a country without adequate visa (or remaining in it after her/his visa expires), and if s/he does not hold the status of "political refugee"' (Chiuri, Giuseppe and Giovanni 2005).

Geo-demographic profile

Regionalization, as discussed earlier, is considered a significant aspect in geo-demographic profiling. Bangladesh, with 40.7 million in its labour force, 46 percent of which are unemployed, has been exporting labour to overseas labour markets for some time. However, there are differences in the volume of migrants across different districts in Bangladesh: not every district sends migrants overseas to the same extent.

In order to examine this variation across regions, data were collapsed into three distinct regions[21] for the BRHKs: i) Region 1; ii) Region 2; and iii) Region 3. Regionalization, however, was not performed for the case of the BRMs as they come from a wide variety of districts, scattered throughout the country, with no obvious pattern for categorization.

The data show that respondents came from 14 districts: Chandpur, Barsial, Patuakhali, Gazipur, Dhaka, Chittagong, Noakhali, Munshiganj, Mymensingh, Bhola, Narayanganj, Sylhet, Sirajganj and Comilla. The respondents from Region 1 came from three districts (i.e. Barisal Sadar, Bhola and Patuakhali). As indicated below, the highest percentage of BRHKs came from Region 2 (Chittagong division), which is composed of four districts. The possible explanation for this lies in the strong regional networking maintained among the population of this area. This type of networking prompts and perpetuates the flow of migration (Chitose 2001). Respondents from Region 3 came from five districts (i.e., Dhaka, Gazipur, Munshiganj, Mymensingh and Narayanganj). In the case of Region 3, two factors explain this phenomenon: First of all, citizens of Dhaka, the capital city of Bangladesh, are more exposed to information flow than in most other districts, and secondly, city-based networking is dense, and powerful in spurring migration. The remaining two districts are in two remote rural parts of the country.

As discussed above, the regionalization of migration, when analyzed as a geo-demographic profile, demonstrates the level of existent network linkages that promote and affect migration patterns. Regionalization also indicates the intensity of migration trends in a particular region. In the case of the BRMs, respondents came from districts that were too scattered to categorize into regions, therefore regionalization analysis has not been performed, as was done in the case of the BRHKs. Unlike with the case of Hong Kong, respondents presently in Malaysia came from the major nine districts, namely: Noakhali, Barisal, Comilla, Dinajpur, Mymensingh, Jessore, Dhaka,[22] Manikganj and Shariatur. These districts are

---

21   Regions have been categorized based on the proximity of the districts to each other within a division. Divisions are composed of a number of districts. The locations of the particular districts are shown in the map.

22   In this division, 3.8 million (9.4 percent) of the population (of equal genders) are economically active (BBS 2004). Out of 11 districts in the Chittagong division, respondents came from four districts, namely, Chandpur, Chittagong, Comilla and Noakhali. The Chittagong division has an economically active population of 7.2 million (18.2 percent).

popularly known for their propensity of migration, and Dinajpur is a district characterized by widespread poverty (BBS 2001).

Interviews with the BRMs ascertained that there are a number of migrants from an area named Zazeera, a sub-district of Shariatpur. This sub-district is known to offer shelter to the migrant workers from their region when they have moved illegally and have lost their travel documents or experienced an expiry of entry. In addition, the authorities in this district may help in obtaining employment abroad for their migrant workers. My interviews with migrants from Zazeera revealed that there are large groups of illegal workers from this district in both Singapore and Malaysia. Their responses gave a vivid example of the importance of regional linkages to the country of destination:

> ... I came here because we knew many people were there from our village. I knew if I had any problem, there was someone to extend me help. (Rajib, a migrant in Malaysia)

Like Mr Rajib, many respondents shared similar experiences during their interviews. Clearly, a regional link cannot be ignored as an analysis factor. Further illustrates was that 23 percent of BRMs came from Noakhali district, followed by Barisal district (19 percent). Eleven percent came from Comilla district, 13 percent from Shariatpur district, four percent from Mymensingh district, 16 percent from Dinajpur district, 9 percent from Jessore district, four percent from Dhaka district and the remaining one percent came from Manikganj district. Evidently, both sets of respondents (BRHKs and the BRMs) came from highly similar socio-economic and demographic backgrounds. However, the two destinations (Hong Kong and Malaysia) are heterogeneous in terms of economic parity, geographical locations and population density.

After arrival to the destination country the primary concern is the legal status of the migrant workers. In order to encompass the range of living issues, this section analyzes the length of stay, legal position, and stay permit, experienced or attained by the respondents.

The data show that a vast majority of the migrants in the BRHK sample had been in Hong Kong between one and six years (71.42 percent). A quarter of them had been residing there for 7–12 years. The mean length of the stay for the BRHKs was approximately five years. In comparison, in Malaysia the majority of the BRMs (51.43 percent) stayed between 10–12 years and thirty percent from 7–9 years with the mean period of stay being 8.93 years. Statistical analysis further shows that there is a significant difference in mean length of stay between the BRHKs and the BRMs (p=0.000).

This data indicates that most of the BRMs are not returning home, even after their work contracts expire. They are staying in Malaysia and thus reporting lengthier stints. However, this does not necessarily mean that Malaysia is a more conducive place to stay for longer period of time: Constrained by lack of money and the expiration of their travel documents, a large proportion of the migrants cannot go back home despite the fact that many of them expressed wanting to. Therefore, the prediction by the Assimilation Theory (more in Part II), that length of stay has a significant relationship with the extent of adaptation to the host society is valid (Valdez 1999). Evidence in support of an adaptation prediction found in this analysis is that, despite the many adversities that BRMs encountered in their host society, their length of stay is quite long.

Table 3.4 Length of stay in Hong Kong and Malaysia

Years (in total)	Hong Kong		Malaysia	
	n=56	%	n=70	%
<3	20	35.71	2	2.86
4–6	20	35.71	21	30.0
7–9	14	25.0	7	10.0
10–12	2	3.57	36	51.43
13–15	–	–	4	5.71
Mean	4.84		8.93	
SD	2.601		3.209	
Significance (HK vs Malaysia)		P=0.000		
Total	**56**	**100.0**	**70**	**100.0**

Note: '–' indicates not applicable.
Source: Author's field data, 2004–2006.

This section attempts to evaluate the number of respondents who were staying and/or working legally abroad and how they managed to do this. Both the Hong Kong and the Malaysian governments have strong policies against illegal workers and those who employ them. Both employers and workers are well aware of these policies. Legal status has much to do with the rationalization process because this largely determines the extent of the migrants' satisfaction on staying abroad and an illegal visa status is perhaps the most dominant factor in migrant worker vulnerability. This book, therefore, addresses the situations of both legal and illegal migrants, drawn in a random sample. There has been a growth in illegal migration worldwide over the last 50 years. Irregular migrants account for an estimated 30 to 40 percent of the 5 million migrants in Asia (IOM 2008, Wickramasekera 1996). However, they often end up in abusive and exploitative situations since they move with little or no information, and remain unprotected in the complex and unregulated circumstances of their journey (Siddiqui 2004).

Since the mid-1970s, as increasing numbers of Bangladeshi workers have sought overseas work, Hong Kong has been a prime destination, due to the ease of acquiring a tourist visa. After the initial tourist visa expires, some are able to arrange for a visa extension (still only as tourists), while others overstay illegally. All of the BRHKs in this study were allowed ingress to Hong Kong as tourists for their first entry. Over time, a few of them managed to obtain work permit. However, at the time of the study for this book, the majority were still carrying a tourist visa and over one-third of respondents did not hold an entry permit at all, while a few others did not have any travel documents whatsoever.

An attempt was made to further explore whether legal position had any impact on respondents' preferences in terms of their work i.e. whether they found satisfaction in working or staying in the destination country while illegal. The question of satisfaction relates very much to the issue of legal status. Although, one important societal factor in adaptation and integration is legal status, it would not be fair to assume that legal status alone speeds up an individual's integration (Piper 2005). In order to explore this, a cross tabulation was performed which shows that most of the BRHKs opted to be 'single illegals',[23] staying in Hong Kong legally, while they worked illegally. The minority elected to be 'doubly illegal', both staying and working illegally, while a few others opted not to respond.

During the interview period, around 36 percent of the BRMs held tourist visas, one-third (33 percent) held no visa, a few others (9 percent) had neither visa nor any travel document (passport), and 7 percent held student permits with the mean length of their stint in Malaysia being 8.93 years.[24] Ambiguous information was discovered, as many respondents claimed to be staying on a valid visa while many others reported that most of these visas were fake. Respondents bought these fake visas at a cost of between RM500 and 3,000 depending on the urgency and category. It is relatively easy to obtain a fake visa sticker because many individuals and private sponsors are engaged in visa selling in both destination countries (Nasra 2002). Therefore, it remains a challenge for researchers to precisely determine whether respondents are staying on genuine visa or on a fake.

Additional tourist permits for Malaysia are only offered to a certain category of people who have access to sufficient funds. In order to circumvent this, many respondents sought admission to private academic institutions (mostly IT) in the country in order to obtain student visas.[25] By paying the tuition fee to private

---

23   The term 'double illegal' refers to migrants' illegal status in both staying and working, while 'single illegal' refers to staying legally and working illegally. Unable or unwilling to reform the systemic problems which contribute to undocumented migration, the government of Hong Kong has taken an ad hoc approach, ostensibly cracking down on undocumented migrant workers while recruiting additional migrants to take their place; meanwhile, employers, police, immigration officials and corrupt agents commit human rights abuses at all stages of the process. The crackdowns are an integrated, multi-agency approach which was launched in February 2002 and led to the deportation of 200,000 undocumented migrant workers (Asian Migrant Centre 2004).

24   Standard deviation is 3.209, for details see Ullah 2005.

25   It is known that it is not difficult to get admission into those institutions if the applicant has at least 10 years of schooling and can afford to pay the tuition fee for the first semester.

colleges instead of paying the agencies or brokers, some migrants are able to get student visas. I interviewed [Bangladeshi] construction workers at Mines area (Extra Super Market) in Ampang, Malaysia who held student visas, however, they never attended the colleges and sought employment instead. They were the 're-returned migrants'. NTV channel in Bangladesh broadcasts a series of reports on this situation (*Jaijaidin* 2006). This situation has come to the notice of the Malaysian government who, upon investigation, found that during the previous year only approximately four percent of graduating students (19 out of 531) returned to their home countries after their degree was completed (*New Strait Times* 2005). This issue of 'missing students' came to light publicly in April 2005.

Table 3.5 Type of permits held

	Hong Kong				Malaysia			
Type of permit	Current status		Status on entry		Current status		Status on entry	
	n=56	%	n=56	%	n=70	%	n=70	%
Tourist	34	60.71	56	100.0	25	35.7	14	20.0
Work permit	2	3.57	–	–	10	14.3	10	14.28
No visa	17	30.36	–	–	23	32.9	42	60.0
No visa, no passport	3	5.36	–	–	6	8.6	–	–
Student permit	–	–	–	–	5	7.1	3	4.28
Business permit	–	–	–	–	1	1.4	1	1.43
Total	**56**	**100.0**	**56**	**100.0**	**70**	**100.0**	**70**	**100.0**

Note: '–' indicates not applicable.
Source: Author's field data, 2004–2006.

Obtaining an entry permit to any country is, for a Bangladeshi citizen, one of the most cumbersome of the formalities related to migration. For example, without a job offer, an offer of admission to an academic institution or an impressive bank balance, Bangladeshi citizens are normally not eligible to apply for a permit to enter developed countries, such as the USA, UK, and Canada.[26] This severely restricts potential migrants' movements to these countries. However, no such formalities are required for Bangladeshis to obtain a permit to enter Hong Kong. As tourists, they are not legally eligible to take up any paid or unpaid employment in Hong

---

26 The average success rate of obtaining the permits is around 10 percent for the USA, 15 percent for UK and for Canada, as stated by an official of the Chanceries of those countries to Bangladesh. Similar patterns are found for Australia, New Zealand, Germany and other European countries.

Kong, however this does not appear to be closely monitored. The table below shows the length of respondents' stay in Hong Kong by the type of visa held. The data yields interesting results in that only three of the migrants interviewed did not possess travel documents; an overwhelming majority of the BRHKs held tourist visas. The respondents stated that there are many people in Hong Kong who recruit and employ illegal workers as they know they can retain them at lower wage than nationals, and also because illegal workers normally accept all types of job, including those that many nationals refuse to do. When asked how he managed a job while staying illegally, one migrant said:

> ... there is nothing wrong with us, Hong Kong people employ us, they violate their own law at first, and then they facilitate us in violating [the laws].

As illustrated in the table around half of the BRHKs prolonged their stay between one to five months. A significant percentage of them stayed between six and ten months and one-third stayed from one to five years.

Table 3.6 Length of stay in Hong Kong in the last trip by type of visa

Length of stay (in months)	Status of visa/permit		Total
	Tourist	**No visa, no passport**	
1–5	25 (44.64)	–	25
6–10	10 (17.86)	–	10
11– 15	6 (10.71)	–	6
20–25	3 (5.36)	–	3
35–40	3 (5.36)	–	3
45–50	3 (5.36)	3 (5.36)	6
55–60	3 (5.36)	–	3
Mean length	15.84		
Total	**53 (94.64)**	**3 (5.36)**	**56 (1000.0)**

Note: '–' indicates not applicable; figures in the parentheses indicate percentages.
Source: Author's field data, 2004–2006.

As mentioned above, the BRHKs could obtain a tourist visa for short period, usually two weeks, upon arrival at the airport. Many Bangladeshi visitors along with other tourists capitalize on this policy. They extend their stay by obtaining visa extensions or by re-entering from neighbouring countries.[27] At the time of the

---

27   In Shenzhen, China, there are many cheap restaurants and hotels. These restaurants are preferred by migrants due to the cheap price and indigenous food. In addition, the

interview, only 37 of the BRHKs held visas that were valid, with varying durations. The validity of the visa was important to each of the migrants because they could be deported and denied re-entry into Hong Kong if they overstayed their visa. Overstayers are detained and repatriated immediately if they are apprehended. If this happens, the only alternative for re-entry into Hong Kong would be to obtain another Bangladesh passport. This is an expensive option, and is avoided whenever possible. If a replacement passport is required, the process involves going home to Bangladesh, and then returning to Hong Kong or seeking a fake passport.[28]

Most migrants extend their visas by visiting Shenzhen via the Lo Wu border station, the nearest and cheapest option for them. Travel to Shenzhen has become a routine part of the migrants' lives. To facilitate the process, they usually obtain a Chinese multiple entry visa, which is obtained by producing fake documents indicating business organizations and business deals in China. When they return to Hong Kong they hope for another tourist visa, and feel 'lucky' if they are given a two-week permit. Keeping their passport and visa in order is thus a central concern for migrants. Despite the expressed significance of the visa, at the time of the interview, 17 of the respondents' entry permits had expired. Thirty percent of the BRHKs had 10–12 days of validity on their permit, 20 percent had 1–3 days and 19 percent had 4–6 days of stay permitted in Hong Kong. Similarly, around half of the BRMs claimed their permit was valid, however the highest percentage of the permits was valid for less than one month. Furthermore, the permits of nearly half of the BRMs had already expired at the time of the interview and majority of them expired over a year earlier. These patterns indicate the complexity and the perseverance of the Bangladeshi migrants in obtaining employment, and indicate the severity of the push factors that are likely to be driving them.

---

transport cost is relatively lower than going to Macao or elsewhere. The restaurants are often managed by Bangladeshi, Pakistani and Indian people, and cater to these migrant populations.

28   *Shaptahik2000* (2005) reports that anyone can obtain an illegal passport from the Bangladesh passport office for a bribe of Tk. 15,000.

PART II
The Migration Process

Chapter 4
Networks and Routes Used to Get to Destinations

The preceding chapter has analyzed the basic socio-economic and demographic profiles of the respondents. This chapter addresses how and why these respondents got to their particular destinations and the ways in which different networks are involved in the migration process. This is particularly significant in the rationalization process of the respondents, because the nature of these networks largely determines the cost and the extent of risk involved in migration, and indicates how easy or difficult the migration would be (Ullah 2010, Rahman 2003, Ainsaar 2004). According to Light et al. (1990) networks promote migratory flows for two reasons: First, network connections support immigration. This support arises from the reduced social, economic, and emotional costs of migration. That is, network-supported migrants have vital help in arranging transportation, finding housing and jobs in their place of destination, and effecting a satisfactory personal and emotional adjustment to what is often a difficult situation of cultural marginality. These benefits make migration easier, thus encouraging people to migrate who may otherwise have remained in their home country. Second, according to Massey's assumptions of a risk diversification model, families send their members abroad in order to minimize risk. Economically precarious households often face high-risk to their well-being, therefore, in the absence of ways to insure against risks of drought, crop failure and natural disasters, migration minimizes overall family income risk (Massey 1989).

Massey (1988) defines migration networks as 'sets of interpersonal ties that link migrants, former migrants, and non-migrants in origin and destination areas through the bonds of kinship, friendship, and shared community origin'. Networks are known to reduce the cost and risks associated with the migration process. Both empirically and theoretically, cost and risk are key factors in making migration decisions. Most migrant workers were found to rely on assistance extended by both recruiting and interpersonal networks for their migration. Recruiting networks were found to be used more to a greater degree by the BRMs than by the BRHKs. This is due to the fact that travelling to Hong Kong is easier in terms of visa endorsement. However, most of the BRMs suffered varied forms of severe adversity *en route* to their destination. Therefore, this chapter also challenges key notions in the literature of networks as facilitating and cost-reducing, because I argue that recruiting networks often put potential migrant workers in serious risk and result in huge monetary and psychological costs.

In effect, networks shape the decisions made by migrant workers about their routes and destinations. The geo-demographic section of the preceding chapter shows that Bangladeshi migrants originate from relatively few districts in Bangladesh, indicating the influence of networks on these particular districts (Rahman 2004).

Types and scales of networks

In this section, I focus on two aspects of networks: their types and their roles. This section further explains how different networks operate at a variety of levels. Specifically, I explain the differences between interpersonal networks and recruiting networks, and between networks that function at different scales – community, national and regional. The term 'network' has a broad connotation; however, this particular research considers networks only in terms of their role in the migration process. It further investigates how networks influence migration decisions by providing relevant information, or misinformation, to potential migrants.

Interpersonal networks are made up of communication pathways that link migrants with relatives, neighbours and friends, abroad or returned. Recruiting networks are comprised of agents, brokers, and those paid to facilitate migration. Often called 'professional' networks, these agents receive money for their services and may operate legally or illegally. They provide transport, visas, labour contracts, housing, legal, extralegal, and other services to facilitate international migration (Alt 2005, Massey, Arango and Hugo 1996, Waldinger 1997, Lindquist 1993, Faist 2000, Jandl 2005). Interpersonal networks are usually not complete alternatives to 'recruiting networks'. They may provide financial assistance, information or accommodation after arrival but are not able to access other services, such as getting contracts and paperwork done. Therefore, this chapter has two sections: The first concerning recruiting networks and their structure and roles; and the second analyzing interpersonal networks and their respective structure and roles.

The presence of recruiting networks indicates that the government, or a registered agent, interacts with and facilitates the labour migration process. In Bangladesh a potential overseas job seeker must register his or her name with BMET (Bureau of Manpower, Employment and Training) or a registered agent. Overseas job seekers must also register themselves with the nearest District Employment and Manpower Office (DEMO). Each registered employee is supplied a Registration ID Card and number to confirm his identity during employment within the country. This registration process is mandatory for all prospective overseas job seekers. In addition, every recruiting agent must be registered with the BMET head office and each is supplied a unique Recruiting Agent ID number. Recruiting agents use this ID number for further queries, and to process or obtain recruitment permission or clearance from BMET (BMET 2008).

Many recruiting network systems have emerged to promote migration in recent years in Bangladesh. The overseas demand for Bangladeshi workers is being managed primarily through three channels: the government, private agencies and relatives of individuals working abroad. A good number of recruiting agencies (RA) has been given recruiting licenses, as per the procedure laid down in the Emigration Ordinance of 1982. Most of these licensed agencies are members of the government-recognized association known as BAIRA (Bangladesh Association of International Recruiting Agencies). Otherwise, there is BOESL (Bangladesh Overseas Employment and Services Limited), another government agency involved in regulating overseas economic migration (BMET 2010). The role of these agencies is increasing rapidly in recent years, due to the widespread unemployment which breeds intending migrants (*Shaptahik2000* 2005).

Recruiting networks are found to work at two levels: Dhaka-based and community-based. These are often inter-connected i.e. community-based networks are linked to the Dhaka-based networks, usually taking the form of outreach networks as opposed to stand-alone agents. Although a community is defined as a unit of social and economic activities (FAO 1983), or a locality – a distinct population cluster – this study considers this concept to embody a position in the hierarchy of the administrative unit. The study demonstrates that both agents and brokers are operating at both Dhaka and community levels. Recruitment at the Dhaka level tends to be more formal, however, at the community level recruitment is often conducted verbally, often even with payments made without a receipt.

At both levels there are two key players primarily active in the migration process – recruiting agents and brokers (popularly known as *dalal*). Almost all recruiting agencies are based in the capital city, Dhaka. The agencies recruit through a host of informal agents and sub-agents, who perform two key functions: the recruitment of workers, and financial transactions. As discussed above, recruiting agents are normally the registered entity with the BMET or with the body of the association of travel agents. The *dalals* (brokers) are not formally registered with either the government or the association of the recruiting agencies, and do not possess any formal identification documents. Brokers often work in an individual capacity, mostly as illegal entities. Brokers often work on behalf of an agent. While there are some unregistered agents, brokers are not registered entities at all. From the information gathered in this study's interviews it became clear that some agents also play the role of the broker. It is interesting to note here that the recruiting agencies do not use the Bengali term for 'recruitment'. They normally use the term '*lok pathai*', meaning 'we send people abroad'.[1]

While around 14 percent of the BRHKs and 23 percent of the BRMs reported having received support from recruiting agents at the community level, 36 percent of the BRHKs and 73 percent of the BRMs were supported by brokers working at the same level. They coerced potential migrants mostly from rural areas with promises

---

1   This might have an inner meaning like helping people. However, to me, it has another meaning. The formal Bengali translation of 'broker' is *dalal* or intermediaries. However, it has another colloquial meaning which implies human trafficker or *adam bepari* i.e. he who deals in human business. Therefore, they might use another term in order to veil these other connotations.

of worthwhile jobs and higher income opportunities (Pasha 2004, *Daily Prothom Alo* 2005). In some cases, agents try to induce the family decision maker, such as a father, grandfather or elder brother, to send a potential member abroad by offering promises of better jobs and higher incomes. According to Bangladeshi culture, the eldest male tends to make decisions for the joint family.[2] Mostly, this position is passed on to the eldest son when he becomes economically more important in the household (Karn 2006). Once the decision is made to send a household member overseas, that family member is handed over to agents who sort out the formalities, such as getting passports issued, signing preliminary contracts, if any, and all other travel procedures. Here it is clear, regarding the proposition of household strategy theory that migration decisions are made collectively at the household level, that this is not applicable to all cases. Before the final move, a 'brokers' session' – a question-answer session on immigration formalities – is arranged. Brokers and agents give a talk on how to perform immigration formalities.

Data show that approximately 83 percent of the BRHKs, and a similar percentage of the BRMs, received assistance from recruiting agents at the Dhaka-based level. Thirty-seven percent of the BRHKs and 70 percent of the BRMs sought assistance from the Dhaka-based brokers. The BRMs appear to be more dependent than the BRHKs on informal brokers, whether they are Dhaka- or community-based, for migration. It is important to note here that it is not the case that those who sought assistance from community-based networks did not seek assistance from Dhaka-based networks. Rather migrants were often handed over by the community-based networks – usually their first point of contact – to the Dhaka-based networks; thus they were often transitioned from brokers to recruiting agents.

Table 4.1 Recruiting networks

Structure of recruiting networks	Hong Kong		Malaysia	
	n=56	%	n=70	%
Community-based network				
Agent	8	14.29	16	22.86
Brokers	20	35.71	51	72.86
Dhaka-based network				
Agent	46	82.4	57	81.43
Brokers	21	37.5	49	70.0
Total	**95**		**173**	

Source: Author's field data, 2004–2006.

2   A joint family is a kind of extended family composed of parents, their children, and the children's spouses and offspring in one household.

Hardships and network

This section deals with the network and routes through which the BRHKs and the BRMs travelled to get to their destinations. Also discussed are the length of time required to get to Hong Kong and Malaysia, and the hardships that participants suffered. Interviews with respondents provided two distinctive pictures of the routes taken: The Dhaka–Hong Kong route has been historically very simple, with less processing time required. Therefore, the role of recruiting networks in arranging for documents and transportation in the case of the BRHKs was minimal. Before 1997, Bangladeshi citizens were allowed to enter Hong Kong without obtaining a visa prior to their departure and were allowed to stay three months. After 1997, a revised entry visa/permit requirement came into force and for Bangladeshis the duration of stay permit has been reduced to two weeks. Again, on 11 December 2006, the policy for entry permits changed. Under the new policy, Bangladeshis are not given any permit upon arrival at the airport: they have to obtain the visa/permit before they fly. This means that their entry into Hong Kong is no longer as easy as it was before. Therefore, potential migrants to Hong Kong are having to rely more on the recruiting networks than in the past. Even so, it is not necessary to obtain contracts or job offers in order for migrants to enter Hong Kong. Therefore, there is still minimal involvement of agencies or government intervention in the process of their migration. In comparison, the Dhaka–Malaysia route has historically been much more complex. The legal migration process is time-intensive for recruiters, involving convoluted visa requirements, and extensive paperwork. However, the illegal trajectories are relatively simpler for the recruiters, involving little-to-no paperwork, but they involve a number of hazards and risks for the migrants (Papadopoulou 2005). Recruiters thus benefit from using illegal trajectories, but at the expense of the migrants.

To most of the BRHKs, the journey to Hong Kong was not a troublesome process. They only encountered minor problems with immigration at the airport in Bangladesh, but most of them were able to bribe the immigration officers at Dhaka to allow them to travel abroad. It is worth mentioning that Bangladeshis do not legally need permission to travel abroad, however, the immigration officers often hassle outgoing migrants by asking unnecessary questions. The migrants reported that the immigration officers asked questions in order to indirectly ask for money. Therefore, this potential barrier was easily overcome, and 73 percent of the BRHKs arrived in Hong Kong directly from Dhaka by standard airline flight. The remaining 27 percent entered Hong Kong through other countries where they had been working previously. Eleven percent came from Malaysia, 5 percent from Taiwan, and another 11 percent from South Korea.

In contrast, most of the BRMs got to Malaysia by passing through a number of transit points in Thailand where they were transferred from one to another group of traffickers. While it is not a major labour importing country, Thailand does import a modest number of workers from Burma and Laos. However, the number of illegal workers arrested in Thailand was as high as 444,636 in 2000 with number of immigrants in the same year recorded at 663,776 persons (Chalamwong 2001). This indicates that Thailand is widely used as a transit country for illegal migration. As much as 10 to 20 percent of Malaysia's total labour force consists of foreign workers (*Migration News* 2005) who sneaked into the country through different transit points with the help of local syndicates. These syndicates are often considered 'mafia-style organizations' (Faist 2000), and are common in Malaysia. They smuggle in foreign labour through certain transit points. Songkhla is a border province in southern Thailand adjoining the state of Kedah in Malaysia. From Narathiwat town, in the province, there are two routes: The first is the Narathiwat-Rangae highway, turning left at Ban Manang Tayo, and continuing on another highway to Amphoe Sungai Padi into Sungai Golok. The second route follows the highway from Narathiwat town to Amphoe Tak Bai, turning right to Highway Tak Bai-Sungai Golok (Forest Department 2005). Migrants gradually gather in the border area, in Hat Yai, Pattani, Songkhla, Yala, Narathiwat and Sungai Golok. When they receive a 'green signal' from the receivers at the end points, they start their journey toward these destinations.

Agents prefer Thailand as a transit country for the migrants that they recruit because of the relatively easier process for obtaining a tourist visa. The majority of the respondents in Malaysia transited through either Thailand or Singapore, with the help of agents or brokers, in order to get to Malaysia. Without a work contract Bangladeshi workers are not given a visa directly to Malaysia. However, in many cases, agents make counterfeit contracts to entice migrant workers. In the period during which Malaysia stopped legally importing Bangladeshi workers, many unregistered recruiting agents emerged to induce migrants using these fake contracts. A tourist visa is not normally granted to Bangladeshi migrants, as it is generally expected that they will seek work once they are in Malaysia. However, under package tour programmes some agents manage to obtain tourist visas to Thailand and Singapore for the migrants, reporting that it is easy to enter Malaysia from both countries. Six routes were reported for entering Malaysia: Approximately 21 percent of the BRMs entered Malaysia by way of Bangladesh to Bangkok and on to Songkhla, crossing into Malaysia through deep forest, in the back of a car or truck; around 26 percent went from Bangladesh to Hat Yai and on to Sungai Golok, traveling to Malaysia by boat and through the forest and hills; 13 percent moved from Bangladesh to Bangkok, down to Yala and Sungai Golok and on to Malaysia through the forest, using trucks or vans; the remaining 27 percent entered Malaysia by going to Singapore and then crossing into Malaysia via the causeway (Ullah 2010).

Surprisingly, the BRMs reported that they were initially not aware that they were going to Thailand without Malaysian visas. They did not even know what kind of visa was endorsed in their passport until they were on board the aircraft. The brokers withheld the passports from them so that they could not check beforehand. Respondents were promised work visas to Malaysia; but instead were given Thai tourist visas. One migrant said that once the full payment was made, the face of the brokers instantly changed. By this he meant that they would start mistreating them.

None of the 42 respondents who provided information about the routes they took to Malaysia had obtained either work permits or any other visas to enter Malaysia. A number of other respondents, who had entered with legal status, became illegal at a later stage. Many of these reported that they had noticed that their rate of wage was lower than had been agreed upon earlier with the brokers. They protested and sought justice; consequently, their employers declined to employ them and refused to give their passports back, thus making them illegal migrants. Piper (2006: 9) reports a similar observation that often the migrants are not at fault for their illegal status: It may be due to the illegal practices of recruiters or employers, such as arranging illegal entry unbeknown to the migrant, or withholding travel documents or wages, all of which are matters entirely beyond the control of the migrants.

> We were handed over at different points to different groups of people we had never seen before. (Akhtar, a labourer in Johor Bahru)

Table 4.2 Routes travelled to get to the destination

Routes travelled by the BRHKs	n=56	%
Dhaka–Hong Kong	41	73.22
Malaysia–Hong Kong	6	10.71
Taiwan–Hong Kong	3	5.36
South Korea–Hong Kong	6	10.71
Total	*56*	*100.0*
Routes travelled by the BRMs	**n=70**	**%**
Dhaka–Bangkok–Songkhla–through deep forest in the back of car or truck	15	21.4
Dhaka–Hat Yai–Sungai Golok, then by boat and through forest and hills	18	25.7
Dhaka–Bangkok–Yale–Sungai Golok and through forest using trucks or vans	9	12.9
Dhaka–Singapore–Malaysia	19	27.1
Dhaka–Malaysia (with tourist visa)	2	2.9
Dhaka–Malaysia (with work visa)	7	10.0
Total	*70*	*100.00*

Source: Author's field data, 2004–2006.

Malaysia-specific hardship and networks

This section discusses the extent of hardship the migrants suffered *en route* to Malaysia. This section excludes the experiences of the BRHKs as the study found that their route to Hong Kong was comparatively not problematic. This is one very significant difference between the BRHK and the BRM populations. The main distinction here is that the BRHKs required less support from networks compared to the BRMs. The central reason for this was, according to the respondents, that many of them knew that travelling to Hong Kong was not difficult and therefore they intentionally did not seek support. This means that the complex migration process to Malaysia has forced potential migrants to seek support from networks, thus increasing their vulnerability to risk and manipulation (Ullah 2010). The higher education levels and greater experience and knowledge of the outside world among the BRHKs, as compared to the BRMs, may also contribute to their ability and decision to not seek help from networks. Most of the BRMs did not have these resources, and thus were targeted by recruiting networks which directed them to their destination through illegal and clandestine routes. Evidence suggests that these routes are associated with high risk, which may result in catastrophic consequences. Many aspirant Bangladeshi migrants, semi-skilled or unskilled, fall victim to the recruiting networks and spend all of their savings in an attempt to find their way to distant lands and a better future[3] (see Ullah 2005).

This trend is evident worldwide. One example worth presenting is a case from 2003, in which one intending Indian migrant was caught trying to transit from Mainland China to Hong Kong, by hiding in a suitcase, risking his life (*South China Morning Post* 2003). Two other recent tragedies took place on separate occasions in March 2005 during which 24[4] young Bangladeshi migrants died of starvation or went missing, in one case, in the Mediterranean Sea and, in another, in the Sahara Desert. Many others were discovered in hospitals throughout different regions in Africa (*Daily Inqilab* 2005, Ullah 2005, *Shaptahik2000* 2005). They had been trying to sneak into Europe to find work (Islam 2005, Islam et al. 2006, *Daily Prothom Alo* 2005; *Daily Ittefaq* 2005).

The usual journey time from Dhaka to Malaysia by air is around 3.5 hours. However, nearly half of the BRMs took approximately one month to get to Malaysia; around one-third took even more time, some even more than two months. Most of this time was spent in a variety of cheap hotels in the districts of Pratunam and Pahurat in Bangkok, and in Songkhla, while the brokers were explored routes and negotiated with traffickers at the next transit point. Some of the BRMs were kept in hotels in Bangkok for a few days, and then shifted to another hotel, probably

---

3   Some 24 Bangladeshi youths, along with several from other countries, attempted to cross the Mediterranean Sea to Spain on a small engine boat, which lost direction and sailed towards Algeria instead of Spain (*The New Age* 2005). These migrants each paid 7–8 thousand USD to brokers for their migration. After nine days, the boat's stock of fuel, food and drinking water was exhausted and 10 of the passengers died of starvation. Later, an Algerian naval ship rescued the rest of the youths and took them to Central Hospital in Boro, where another youth died (Ullah 2005, Cox 1997, Hamilton 2002).

4   This estimate varies over the sources. However, this number is from the *Shaptahik2000* (2005), a source which covered these tragedies thoroughly, including conducting interviews with the survivors.

to avoid suspicion on the part of the police and immigration officials. The brokers continually reassured them that they would shortly start their journey to Malaysia and transported them to Songkhla. Often, though, this would be a false start, and they would be moved back to Bangkok again, where they would stay another few weeks, before returning to Songkhla, and then on to Sungai Golok. For many of the migrants (12 percent), this process was repeated several times; they finally set off for Malaysia two months after their departure from Dhaka. The exceptionally long time taken for their travel is only one added dimension of the vulnerabilities experienced by migrant workers. Other common vulnerabilities that migrant workers encounter are contract substitutions, excessive fees, and promise of non-existent jobs by unscrupulous recruiters (Capulong 2001). However, it is astonishing that migrant workers spend between one and three months on their journey that should take only four hours. The remaining 28 BRMs (40 percent) managed to get to Malaysia in one day. They reported that this was possible because some of them had managed to obtain work permits or tourist visas to Malaysia. Therefore, they flew to Malaysia directly from Dhaka by plane.

While migration, even under normal conditions, involves a series of events that can be highly traumatizing and place migrants at risk (Carballo and Nerukar 2001), illegal migrants are consistently more exposed to vulnerabilities. In this study, the ordeal suffered by the BRMs on their journey is reminiscent of other migration case studies. For example, studies by Waldinger (1997), Waddington (2003), Carballo and Nerukar (2001) and Ullah (2010) show how irregular migrants are prone to a number of varied exploitations. The following comment by one of the BRMs reveals the extent of the hardship experienced on their way from Thailand to Malaysia.

> We, a group of seven youths, were asked to get ready while we were in Songkhla and to start for Malaysia at midnight. Later we were directed to a deep and dark forest. We stepped out without knowing where we were going. (Mujib, a labourer in Kuala Lumpur)

> A group discussion with the respondents involved in this particular journey revealed that they were not allowed to make any noise during the journey and many of them said that they had no opportunity to retreat although they knew that they were going to enter Malaysia illegally. They had to go forward, risking their lives, to an unknown destination. They reported hearing a loud scream from behind as one of the members of the group begged for help; however, no one was allowed to look back. In the morning, when they stopped in the jungle to wait for the night to come again, they found that one of their members was missing. No information about him was ever received. They feared he was bitten by a poisonous snake and that he had died. The brokers paid no attention. (Ullah 2010)

The migrants later heard many stories of this kind from their compatriots. They would travel for many days, living without shelter or water for bathing, and in fear of snakes and poisonous insects, which the group had to endure in silence. River and sea crossings were part of the journey: One of respondents said that 'while on board a small boat in the dead of night on the sea,[5] the feeling was that we could not arrive on the shore alive'. The migrants were forced to get out of the boat when it was still 300–400 metres away from the shore.

> We swam ashore. It was like we were being killed by the traffickers. We had to
> swim to reach the shore with one hand and had to save our belongings with the
> other. (Ratan, a labourer in Kajang, Malaysia)

Respondents, while narrating the hardships they suffered *en route* to Malaysia, reported that they had regarded it as a life-threatening venture. Around 19 percent said that snakes or poisonous insects bit them, however they survived by applying indigenous treatments they knew from back home.[6] Around 59 percent were bitten by large leeches, while 22 percent were injured by running into trees or stumbling on their way at night. Many others suffered dehydration and some suffered fever and jaundice. A majority reported weaknesses, due to the lack of sufficient food and to mental stress, fear and uncertainty.

Migrants also suffered from the lack of food supplied to them along the way; they travelled for days with limited or no food. The study data show that around 21 percent went without food for two days on the way to Malaysia from Songkhla, due to incorrect estimates made by the brokers about the length of the trip. They took dry food for only five days, however the journey took more than 7 days, and the food ran out after four days. Around 46 percent survived for two days by drinking only water, while 33 percent ate insufficient quantities of biscuits and bread for three days. If they asked for more water or food, the brokers pretended that they did not understand their language and remained indifferent (Ullah 2010).

According to the key propositions of networks theory, migration networks are significant in the process of migration because they reduce the cost and risk of migration (Massey 1989, Light et al. 1990). However, as evidenced by the current research, the volatile combination of unprincipled recruiters, desperate would-be workers, lax government oversight, and corrupt immigration officers leads to lengthy, painful, dangerous, and expensive journeys for migrants. If we look at the recruitment procedure closely, we can see that the vulnerabilities of migrant workers are embedded in the system itself: they are rooted in the policy of the Bangladeshi government in that it gives the recruiters a near-monopoly in organizing labour migration. In exploring the risks borne by the migrant workers, the findings clearly demonstrate that the role of the recruiting network stops abruptly after the migrants' arrival in the destination country. Therefore, their role is limited: These 'professional networks' do not deal with or ameliorate

---

5   Strait of Malacca or Gulf of Thailand.
6   This means applying a paste of leaves (made by masticating) which have medicinal value on the wound for healing.

any of the risks migrant workers face while they are abroad; here-in lays a serious shortcoming of network theory. I argue that with time, the importance of networks will decrease as people gain better access to information on world labour markets and recruitment procedures due to technological development. Network theorists share a general tendency to ignore self-induced migration. If we look closely at this study's findings, it is evident that an overwhelming majority of the BRMs had to face increased risk for the duration of the migration process due to relying on professional networks. Moreover, the study did not find any evidence that use of these networks reduced the cost of migration. The following chapter demonstrates that most of the migrant workers (both BRHKs and BRMs) had to spend exorbitant amounts of money in order to finance their migration. This chapter further demonstrates that those who did not rely on professional networks spent relatively low amounts as compared with those who relied on the networks. Therefore, the preposition by network theory of network use reducing cost and risk is clearly not applicable to the situation of Bangladeshi migrant labour. Hence, the general concept of the theory has been challenged by empirical data.

The decision to migrate may be made after due consideration of all available and relevant information, rationally calculated to maximize net advantage, including both material and symbolic rewards (Middleton 2005). This information is gathered from multiple sources, and while the media's role in transmitting information is crucial for migration, the role of additional sources of information cannot be ignored. Specific information on particular issues and possible destinations is normally received from the agencies or brokers. The following section discusses what the precise information was that the study respondents received that influenced their decision to migrate. In this study, the majority of respondents claimed to have gathered information about the socio-economic and cultural setting of the country of destination before they decided to migrate. For most, the kind of information they received before their move helped them to make up their mind about migration. Further findings were illuminated by the present research. Prior to their arrival, one-third of the BRHKs had planned to enter Europe, using Hong Kong as a transit point, and around 36 percent had the idea that Hong Kong would be a rich and 'good' country for finding work. A few of the BRHKs reported believing that Hong Kong is a country close to Bangladesh. Several also reported that they decided to go to Hong Kong because getting there 'does not cost much money'. Many of the respondents came first on a 'test' basis, i.e., they could go back home if they could not find work. Approximately 38 percent knew prior to arrival that the income to be earned in Hong Kong would be very high. Around half of the BRMs (46 percent) were told that Malaysia was a country where they could obtain work easily with high-income probability. A few others were told that Malaysia was a rich and 'good' country (17 percent); therefore, they agreed to migrate, while the remaining respondents chose Malaysia as their destination country because it was a 'Muslim' country (6 percent). As mentioned the role of the media is important in shaping the decision of potential migrants. Currently in Bangladesh media is playing a major role in promoting migration, and making the process safer. Both electronic and print media sponsored by the government warn potential migrants not to depend on illegal recruiters for their migration. However, this initiative only started only recently, so its effectiveness has yet to be fully evaluated. Therefore, I agree with the respondents of the study that their migration decision was not primarily a media influenced or facilitated venture.

A significant difference is observed in this study between BRHK and BRM respondents' perception of the accuracy of the information they received prior to migration (see Figure 4.1). The data show that a majority of the BRHKs said that the information they received about the country and the income level before leaving Bangladesh was correct. Figure 4.1 illustrates that a significantly higher percentage of the BRMs thought that the information they received was not correct as compared with the BRHKs. They evaluate whether the information they received was correct or not based on the level of hardship and deprivation they suffer during the migration process and the problems they faced in Malaysia, which are reported to be much worse than in Hong Kong.

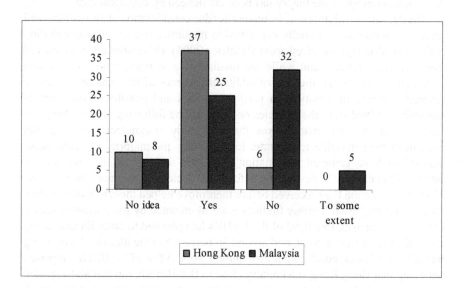

Figure 4.1 Accuracy of information received prior to migration
Note: The units in the graph represent percentages.
Source: Survey 2004–2006.

Many of the BRMs also recalled the ways in which they were deceived by the brokers (as will be illustrated in the subsequent sections of this chapter) and that they were not told ahead of time about the routes they would take to get to Malaysia. Many migrants added that they were so influenced by the information that they made the decision to migrate even though in order to acquire finances for migration they had to sell their assets, and/or become indebted to banks, NGOs, informal moneylenders, relatives, friends and neighbours. Only a few respondents said that they did not pay heed to the promises given to them by the agents as they believed that they normally exaggerate information to attract clients. Household members also believed those promises and were encouraged to contribute in order to send them abroad. A few of the respondents said that 'now they deliberately forget what they heard from these agents and brokers', because reminders of their false promises cannot give any solutions to their existing problems and is therefore only frustrating. As a result, many of the respondents were reluctant to reconsider past events. For example, Anowar, a migrant worker in Hong Kong, said,

> ... now it is not a time to think about the accuracy of the information we received [before migration]. We [have] already moved [so] we have to adjust to the existing situation; there is no use to think about the past.

Interpersonal migration networks are usually interpreted as interpersonal connections that link migrants, former migrants and non-migrants between home country and host country through ties of kinship and friendship (Massey 1989, Faist 2000). These pathways offer an important data exchange which work to mitigate imperfect information about the labour market in the host country. Interpersonal networks have long been significant in the migration process for providing required information and support. Nasra and Menon (1999) found in their studies that 63 percent of the Bangladeshi migrant workers they interviewed had received information about job prospects from friends and relatives overseas, approximately 12 percent had arranged their migration through friends, and about 16 percent had been assisted by friends in financing the cost of migration. This suggests that in Bangladesh society, recruiting networks largely influence migration.

Similarly, this study finds that both the BRHKs and the BRMs depended considerably on interpersonal networks for aspects of their migration. More than half (56 percent) of BRHKs reported having received assistance from their relatives for their migration, more than half (52 percent) from their friends and 16 percent from their neighbours. For the BRMs, around one-third (17.4 percent) received support from their relatives, 7 percent from friends, and 7 percent from neighbours, suggesting a higher dependence of the BRHKs on interpersonal networks than the BRMs.

Kinds of support

A number of ways have been identified as to how interpersonal networks enable others (relatives and friends) to migrate. Reciprocity, whether formal or informal, is fundamental to the operation of these types of networks. When one party receives assistance from another, some form of return is expected. In other words, people help those who have helped them before, in any way (Rahman 2003). My data show that of the 13 BRHKs who received financial assistance during their migration, eight (14 percent) later reciprocated by extending support to their relatives, primarily by financing further migration. For the BRMs, 29 (28 percent) received financial assistance from their relatives and, in return, around 12 offered financial support to help their relatives to migrate. Furthermore, interpersonal networks are also used to transfer remittances. Around 29 percent of the BRHKs and 31 percent of the BRMs used their compatriots (friends, neighbours and relatives) to transfer remittances to Bangladesh.

Table 4.3 Support extended by recruiting networks to respondents

Roles played	Hong Kong		Malaysia	
	n=56	%	n=70	%
Respondents received financial assistance	13	23.21	29	41.42
Respondents received offers of financial assistance	8	14.29	12	17.14
Respondents decided to offer financial support	3	5.36	4	5.71
Respondents received offers of visa support*	–	–	7	10.0
Respondents received visa support*	4	7.14	11	15.71
Channelling respondents' remittances**	14	25.0	32	45.71

Note: '–' indicates not applicable; * Visa support means sponsorship of a migrant by giving a bank statement showing one's bank balance; ** Sending remittances through networks (other than formal channels, such as banks).
Source: Author's field data, 2004–2006.

The respondents had received information from their friends, relatives and neighbours about their country of destination before they migrated. This had given them the opportunity to check the information given by the agents or brokers. Questions might arise when they received information from their relatives and friends as to why were they being deceived. However, this chapter has clearly demonstrated that the BRMs in particular suffered adversity which is often unpredictable *en route* to their destinations. I argue here that these relatives or friends were not aware of the level of suffering because most of those already settled in the host country had migrated under government labour contracts and therefore had not experienced the same, extreme, situations. However, many of the respondents said that friends had promised them many things like help finding a job and cheap accommodation but once the new migrants arrived, they were unable to be very helpful. Respondents reported that they seemed busy with their own work. It is clear that the information received varied from and content between the BRHK and the BRMs. There is also a clear difference in the type of information provided by the recruiting networks and the interpersonal networks. Recruiting networks' information was focused on the higher income and opportunities to sneak into other developed countries while interpersonal networks' information was more practically-oriented.

Table 4.4 Information received from interpersonal networks

Information received	Hong Kong		Malaysia	
	n=56	%	n=70	%
Living is cheap	3	5.36	17	24.29
Easy to find/free accommodation	12	21.43	23	32.86
Easy to get work	19	33.93	44	62.86
Good and rich country	50	89.29	21	31.43
Muslim and friendly country	–	–	16	22.86
Easy to stay longer period	–	–	14	20.0
Income is high	37	66.1	9	12.86
Income is low	2	3.57	29	41.43

Note: '–' indicates not applicable.
Source: Author's field data, 2004–2006.

Chapter 5
The Finance and the Costs of Migration

The chapter provides an account of the financial resources spent on migration by migrant workers. I argue that the process of obtaining these resources greatly influences migrant workers' process of rationalizing their migration decisions. This is because very often migrant workers do not have the financial resources to fund their migration outright. Therefore, they try to obtain financial resources from diverse internal and external (e.g. borrowed funds) sources. In most cases, external sources extend financial support conditionally on a basis of reciprocity, which entails collateral. Internal resources are substantiated by selling their landed property and other productive assets. Apart from paying monetary costs at the time of departure, they continue to pay non-monetary costs in terms of psychological, social and opportunity costs for the duration of their time abroad. Many migrant workers take years to pay off borrowed money. Many others could not buy back their properties that were sold at the time of departure. This means that the process of obtaining financial resources to fund migration has long-term effects which often become a perpetual problem. Therefore, the issue of finance is intimately related to the process of rationalizing the migration decision.

While the main focus of the present discussion is on financial issues, this chapter also analyzes the psychological and social costs of migration. Obtaining the required amount of financial resources involves a number of factors such as: the sources of money, amount borrowed, and amount spent to finance migration, repayment of loans, and the impact of borrowing on the livelihoods of the migrants and their families. Migration involves costs, which range from the initial expenditure for transition to a settlement in the host country, as well as the perceived cost of migration attributed to the psychological toll for families and friends left behind (Chau 1997).

Existing research pays very little attention to the costs of migration, including money channelled out during the departure process. In analyzing cost-benefits of migration, Ruiz (1992), Chau (1997) and Abrar (2002) have highlighted the cost incurred by verses gained through migration. However, to my knowledge, only the IOM (International Organization for Migration) (2001 and 2005) has so far included the issue of money being channelled out of the sending country through the migration process. While labour migration contributes financial resources in the form of remittances to the nation's economy, it also channels a significant amount of money out of the country during the departure process. The costs involved in migration include charges for passports, airfare, visa and medical tests. Although some of these expenses are spent within the originating country, such as passport fees which go towards government revenues, and potentially airfares if state airlines are taken, much of these financial resources are transferred externally, through costs such as, airfares on foreign airlines, visa fees for destination countries, and potentially payment to recruiting agencies based outside of the origin country. In this case, medical tests are done by doctors in Bangladesh; therefore, this cost does not transfer out of the origin country. There are additional personal costs for clothing and other preparatory items. Moreover, the migrants have to pay a sizeable amount of money to the recruiting agents, usually in the form of agency commission (IOM 2002, Siddiqui 2003, Miyan 2003) which constitutes the largest portion of the total cost. This chapter, therefore, explores the degree of influence held by the cost of migration on the migration decision.

Though the opportunity to migrate is limited for the poorest sector of the population, due either to a lack of the necessary capital to finance migration or to lack of required skills (Carling 2005), the long queue of potential migrants for overseas employment and the proliferation of recruitment agencies in Bangladesh clearly indicate the growing interest in labour migration (Rahman 2004). Since labour migration generally takes place in relatively poor households (Karim 2006, Ullah 2006), it is no wonder that potential migrants' access to sufficient financial resources, marketable skills, and information is also limited (see Carling 2005: 2). In particular, they lack the cash-capital to finance migration outright, and have to struggle to obtain the required amount of money, often seeking multiple sources of funding.

In contemporary literature, migration is considered a kind of investment (Abrar 2005). However, the issue of financing migration and its pertinent factors, such as the hardship in obtaining funding, has not received sufficient attention in the literature. Individuals, as rational beings, invest in the schemes which offer more profit (monetary and non-monetary) than other perceived schemes. Many factors, such as age, health, marital status, wealth, and tolerance for risk, are considered in migration decision-making. Investment involves risks, often called the risk-return trade-off. However, usually, the risk and return are positively correlated; therefore, in order to reduce risk one must be willing to accept lower returns. The financial factor plays the most influential role in deciding whether an individual migrates or not. In other words, economic differentials are the most significant determinants of migration (IOM 2005). The decisive benefits then are the expected differences in income between home and host countries, and the monetary costs of transporting the migrant, between the two places. In making a migration decision, a cost analysis is required entailing both monetary and non-monetary costs (Sjaastad 1962).

Conventionally, labour migrants are both pushed by the lack of opportunities in their home country and pulled by their hope for economic gain in the destination country. Migration often involves travel from a region which is poorer, economically, to one which is richer. Thus, I argue that migration costs depend more upon how rich the destination country is and how high the potential wage is, not on how far the countries are apart. Therefore, estimating migration costs on the basis of a given distance between two places, as Sjaastad (1962: 83)[1] did, is not strong or sensitive enough to calculate the whole range of cost components such as social, psychological and opportunity costs. What is crucial in the calculation of migration costs is the difference between expected incomes in the host country vs. the costs of getting there. According to DeJoung and Fawcett (1981), migration intentions flow from the perceived value attached to the outcome of a migration decision, often by the probability that it will lead to the desired outcome. This model then assumes that people make choices that will maximize their outcome and well-being (Golledge and Stimson 1997, Ainsaar 2004, Kephart 2005).[2] However, this mode of cost-analysis tends to motivate recruiting agents and brokers to exaggerate expected incomes to potential migrants, as this would have the effect of balancing, and making more palatable the perceived migration cost. The supposedly rational decision-making by migrants may thus be based on disinformation, leading to unintended outcomes. Along with monetary costs, migration also contributes to the psychological stress, both of the migrants and that of their family members. Although the quantification of psychological or social cost is difficult, it is actually

---

1   In determining the migration cost some researchers intend to include the distance between the destination and the origin country (Sjaastad 1962). Gravity model, though a very influential one in migration study, has been excluded from the framework of the current study mainly because of its argument regarding migration cost. The simplest and most powerful of macro level models is the gravity approach (Ravenstein 1985). This model of migration is concerned with the relation between distance and the propensity to move, arguing that choices are made relative to the distance of the destination, and that people intend to migrate only for short distances. A little more discussion on the gravity model is required to explain my stance on the question of distance and a determinant of migration. According to a gravity model, migration costs are relative to the distance to destination, therefore people only intend to migrate to locations that are at a short distance from their home country. The seven resulting laws of the model are: the majority migrate only short distances and thus establish currents of migration toward larger cities; this causes displacement and development processes in connection with populations in emigration and destination regions; the processes of dispersion and absorption correspond with each other; migration chains develop over time; migration chains lead to external movement toward centres of commerce and industry; urban residents are less prone to migrate than rural people; and the last law, also true for the female population, is that the number of migrant events between two regions is directly proportional to the number of inhabitants in each region and indirectly proportional to the squared distances between the out-migration and the in-migration regions. However, and consistent with some other studies (Ullah 2010 and 2007), the present research shows that many variables determine the cost of migration other than the distance factor.

2   Expectancy-value theories hold that people are goal-oriented beings. The behaviours they perform are in response to their beliefs and values, and are undertaken to achieve some end. According to one of the components of the value expectancy model, migrants might have a set of preference orderings in terms of their decision where to migrate. For example, they might have a number of options, such as a job offer at home or in another country, or a business opportunity, and therefore they choose the best option as perceived by the particular person and sometimes the family.

important to the cost analysis of migration that intending migrants engage in. My interviewees expressed being concerned about how their children, spouses and siblings would be cared for in their absence. All the psychological costs reported in their cost and benefit analysis of migration.

Migration has a positive impact on the development of economies in source countries through easing unemployment, remitting foreign currency for the balance of payment and commodity exports, and augmenting government revenues. However, the cost of migration might impact negatively on the economy of the source country as well, by draining out; draining out skills and knowledge, and incurring social and psychological costs. In this context, due to the complexity of the system, the cost and benefit aspects of migration have been categorized into monetary and non-monetary terms.

This section analyzes the major components of the monetary cost i.e. payments to the recruiting agents in the form of fees. In the initial stages of labour migration during the mid-1970s, recruiting agencies in Bangladesh used to receive a commission from the overseas employers for their services and did not charge the migrants themselves (Miyan 2003). Additionally, the cost of airfare was borne by the overseas employers. However, over the years the arrangements have changed and the burden of all charges and travel expenses are now borne by the workers. This change may have occurred due in part to the fact that many countries now have come forward to export labour to the receiving countries, increasing competition. This has given labour import countries more choices, giving them greater control in imposing conditions on the sending countries. Along with this, many illegal agents emerged to facilitate migration. They often charge exorbitant fees, ignoring all established rules on migration costs. This cost is unusually high, in particular, in Bangladesh, when compared to other migrant sending countries in South and South East Asia. For example, a migrant in Nepal has to spend around US$1,000 (or even less if there is free visa) to finance migration to Malaysia (Samren 2007), however more than one-third of the BRMs spent approximately US$3,000. Prospective migrant workers often get cheated due to the intricacies of the migration process and the long chain of intermediaries involved (see Chapter 4; also see IOM 2003, Miyan 2003).

Although the Government of Bangladesh has fixed the level of commission[3] for the recruiting agents that can be charged, the surreptitious nature of these

---

3   The Bureau of Manpower Employment and Training (BMET) has a number of fees and taxes, which work out at between 2,200 and 3,100 Taka per migrant depending on their destination and skill level. These fees help run training programmes for migrants and contribute to a migrant workers fund that is supposed to be used to help migrants who run into problems in destination countries. Often migrants arrange travel, visas and jobs through agencies who charge fees for their services. The Bangladesh Government sets legal maximums for the amount that agencies can charge migrants. These have been updated recently and made specific for different countries or regions. In the last few years these were supposed to be 8,000 taka (USD 116) for an unskilled migrant and 12,000 taka (USD 174) for a skilled migrant, although it seems that these legal maximums are not always applied.

transactions makes it impossible to enforce. Migrants must pay cash, in addition to the agency fee which is given completely without document. The surreptitious nature of these payments allows recruiters and brokers' fees to be many times higher than the government-determined rate. These fees are even higher for illegal migrants. It is highly common for aspirant migrants to fall prey to illegal recruiting agents and or brokers who extort money from them.

This section shows that the exorbitant cost has caused economic hardship to a large proportion of migrants' families. My respondents (most of whom were illegal) often expressed their frustrations about this. In this section I identify and discuss an important element of this hardship: the excessive financial cost of migration leading potential migrants to borrow large amounts of money from multiple sources. In many cases migrants obtained money by selling their productive assets and many others were deeply indebted to informal moneylenders.

I found that migrants paid an average of between Tk. 7 and 8 lakh (US$12,000–14,000) in order to finance their migration (Ullah 2005).[4] Similar data were found in a study by Islam (2005: 1), while Rahman (2004: 4) estimated that the cost is often so high that it is impossible to recover it by working overseas for two years. In fact, in the last two decades, the benefits of migration for unskilled migrants have dried up considerably (see Rahman 2005: 2) and the benefit of migration often varies depending on the number of contract periods. During the mid-1970s, migrant workers used to earn a salary of USD 300–400 for a 40 hour work week. For similar types of work, at present a worker earns a salary of USD 80–100 per month for a 60 hour work week (Siddiqui 2001). Therefore, the intention to recover the costs incurred by the migration process often leads migrants to stay longer in the destination country, even after their contract expired, eventually leading them to be illegal migrants. The following section discusses the amount of money spent on migration.

As mentioned previously, the amount spent by migrants reflects not only transportation costs to the destination country, but the agent's or broker's fees as well. The cost is not dependent on the distance; rather, it depends on the policies of the sending and host countries; how the recruiting agencies follow or evade the policies; and the desire of migrants themselves to move to a particular country. The general trend in the Bangladesh out-migration scenario is towards a heavy dependence on the professional networks (INSTRAW/IOM 2000), often to the

---

Currently this has been raised to a more realistic figure of 70,000 taka, including airfare, for migrants going to the Middle East (see Hasan 2002). This correlates more closely with the charges found to be paid by migrants in studies such as Siddiqui and Abrar (2003), Afsar (2001a) and Zeitlyn (2006: 26).

4 Carling (2005: 2) says that the cost of being smuggled can go into tens of thousands of US dollars, but it depends on the origin, destination, terms, and modes of transportation. Rahman (2005) indicates that a migrant worker in general needs to spend around US$2,000 to 2,500 for migration to Middle East and South East Asian countries for a two-year contract, if any.

disadvantage of the migrants. Along with the recruiting agencies, many illegal brokers' agencies have emerged to facilitate migration. For both BRHKs and BRMs, many respondents sought support to facilitate migration from both legal and illegal brokers, however, it is clear from the data that BRMs relied more on the brokers for their migration than the BRHKs. Theoretically and empirically, the more the migrants depend on the agents, the higher the cost. Therefore, the BRHKs generally paid less in agency fees. The study further shows that BRHKs spent a significantly lower amount overall (P<0.003), as compared to the BRMs, to finance their migration.

According to most of the respondents, interpersonal networks through friendship, kinship or neighbourhood ties are more helpful than the agencies or brokers in reducing the cost of migration. A number of studies show similar findings: For example, the amount spent on migration shown in the current study parallels the findings by Siddiqui and Abrar (2003).[5] The number of migrants interviewed in these studies, however, is not known.

Table 5.1 Amounts spent for migration

Amount* (Tk.)	Hong Kong		Malaysia	
	n=56	%	n=70	%
<40000 (US$588)	1	1.79	–	–
40000–80000 (US$588–1,176)	35	62.5	21	30.0
80000–120000 (US$1,176–1,765)	17	30.36	23	32.86
120000–200000 (US$1,765–2,941)	3	5.36	26	37.14
Mean	78267.86		109142.86	
Standard deviation	27005.381		30323.992	
Significance (HK vs Malaysia)		P<0.003		
Total	**56**	**100.00**	**70**	**100.0**

Note: '–' indicates not applicable; * During the study period, the exchange rate for US$1 was Tk. 50.
Source: Author's field data, 2004–2006.

---

5   Cost of migration to Malaysia or to the Middle East is around 100,000 Taka (about USD 2,000) (Afsar 2000, Siddiqui and Abrar 2002) which is about equal to Mahmood's estimates in 1996 and 1998. Migration to Singapore costs between Tk. 150,000 and 200,000 (S$2,300–3,100) (see Siddiqui and Abrar 2003), which is significantly less than that Mahmood's (1996) estimate of S$5,000. However, Rahman (2004) in his study found that the majority of his respondents (61 percent) spent between Tk. 180,000 and 250,000 (USD 2,647–3,676).

A focus group discussion with the study respondents was conducted in order to evaluate the breakdown of the costs incurred by migration. The following table gives the breakdown of mean expenses: approximately Tk. 78,267.86 (USD 1,151) for the BRHKs and Tk. 109,142.86 (USD 1,605) for the BRMs. For BRHKs, the airfare has been the main item of expenditure followed by 'cash' given to the agents, then by agency/broker fee.[6] Significantly, the BRMs spent the highest amount on the 'fee' to the agents followed by airfare.

Table 5.2 Itemized breakdown of migration costs

Items of costs	Hong Kong		Malaysia	
	Taka	% (of total)	Taka	% (of total)
Passport	2500	3.20	4000	3.67
Agency/broker fee	10000	12.8.	45000	41.30
Medical tests	No data	No data	2500	2.30
Air fare	33500	42.84	40000	36.60
Clothing/other preparation, luggage	5200	6.65	3000	2.73
Cash given to agents	22000	28.13	12000	11.10
Others	5000	6.38	2500	2.30
Mean costs	**78,200**	**100.00**	**109,000**	**100.00**

Source: Author's field data, 2004–2006.

As indicated in Part III, the respondents' profiles confirmed that they came from relatively poor families, that a great majority of them were living below the poverty line prior to migration, and that most of them were unemployed before they left Bangladesh. This suggests that the majority of these migrants did not have the adequate cash capital to finance their migration. The financial cost of migration is immense in the context of Bangladesh, where the country's GDP per capita was USD 2,011 in 2006 (GoB 2006), and is, in most cases, beyond the capacity of rural families, as indicated in the respondents profiles, where their average yearly income at home was approximately USD 134. Therefore, potential migrants have to raise funds for migration from multiple sources. While there is a noticeable absence in availability of institutional financing to migrant workers in Bangladesh,[7] potential migrants are able to seek out funding from various informal

6 The respondents reported that they got a cash memo for 'agency/broker fee', however, no such memos were given for the 'cash given to agents'. This is how they differentiated these two ways of giving money.

7 Bangladesh will set up an Expatriate Bank soon for the welfare of the expatriate Bangladeshis. The government would also set up two separate wings in the Expatriates

sources in addition to individual/family savings and internal sources[8] (Miyan 2003). Thus sources have necessarily become diverse, as a single source alone, in most cases, cannot support migration.

It is important to clearly define the external and the internal sources of funding as further discussion will draw upon this information. Major internal sources for borrowing money consist of relatives, friends and neighbours, as well as the sale of assets such as jewellery and durable household goods, the mortgaging of land or other assets, and the liquidation of savings. Gifts or loans from family and friends are also common sources of finance. External sources, consisting of formal and informal money-lenders, usually pose conditions of high rates of interest on transactions, as they are not based on collateral: conventionally, in order to secure a loan, banks impose collateral agreements on borrowers e.g. deed or certificates of land ownership. In this case, physical collateral is often unfeasible, therefore collateral is replaced with 'conditions'. The accumulated interest often goes as high as the principal amount, doubling what the migrant has to pay back. Because of this, most migrants use their own savings, if possible, or sell assets such as land, jewellery, durable goods or livestock, although some do take out loans from banks or moneylenders (Afsar 2001a, Zeitlyn 2006). It must be understood that it is not simply the case that the migrants undervalue their own assets in favour of expected monetary income; they are willing to risk the assets they have at hand because, in many cases, migrants believe that initial economic migration can lead to entrance into developed countries where they can obtain citizenship. Therefore, it is not only the expected monetary income they considered.

Data show that a majority of migrants borrowed from external sources to finance their migration. Social capital and kinship ties are extremely important in obtaining funds for migration: Well-off relatives often want their poorer relatives to become better-off and therefore extend help by lending money. Friends finance the migration of other friends with the hope that they will, in turn, help them to migrate at a later stage (Rahman 2003), reflecting a reciprocal and altruistic ideology. A study on Bangladeshi migrant workers in Malaysia found that, in order to finance migration, these forms of borrowing money from external sources was common (Abdul-Aziz 2001). In terms of NGO support, micro-credit programmes do not lend money to migrants directly, however, beneficiaries (who are mostly females) often borrow money from the NGOs and hand it over to their husbands or other relatives, to aid in migration. My data show that a higher percentage of the BRMs, compared to the BRHKs, sold their assets to obtain the funds necessary for migration. However, none of the BRMs borrowed directly from NGOs. The International Organization for Migration (see IOM 2005) indicates that, out of an estimated average cost of Tk. 77,000 (USD 1,132) for migration to Malaysia, the

---

Ministry, namely administrative and planning, with precise focus on particular areas for sending Bangladeshi jobseekers abroad in a 'synchronized fashion' (*Daily Star* 2010).

8   Sources are of two types: Internal i.e., obtaining money by selling one's own assets, and external sources i.e., NGOs, money lenders (formal or informal), friends and relatives.

highest proportion of the funds came from personal and family sources, followed by the sale or mortgage of landed property and then private loans. Researchers found a positive aspect to this form of borrowing money for migration: the diversification of funding sources, from relatives, friends, money lenders, and the sale of family (shared) assets, minimizes the financial risk of migration. The decision to migrate is often made within a household, which suggests that the collective decision to send a family member abroad minimizes the risk of various crises (Batistella 2002). I do not fully agree with Batistella because borrowing from numerous sources leaves an individual or family heavily indebted to many people, both in monetary and non-monetary terms. However, according to the respondents, diversification of funding sources e.g., relatives and friends, allows them to borrow a small amount from each. Therefore, the pressure for paying off the debt quickly is less.

Table 5.3 Sources of finances (multiple response)

Sources	Hong Kong		Malaysia	
	n=56	**%**	**n=70**	**%**
NGOs	6	10.71	No data	No data
Money lenders	10	17.86	22	31.43
Relatives	7	12.5	36	51.43
Friends	6	10.71	11	15.71
Selling assets	26	46.43	38	54.29
Mortgaging off land or other asset	9	16.07	19	27.14
Total	**64**		**126**	

Source: Author's field data, 2004–2006.

A case study shows the process of how a migrant acquired the funds necessary to pay for his migration: Ali, aged 37, worked as a labourer in a textile mill in Malaysia for about five years. From Gazipur district, 10 kilometres from Dhaka, he had 10 years of schooling with no other training or experience. He was the eldest son among four siblings in the family and was married. He was the only family member earning a wage for the family. His income was not adequate for family subsistence. One of his friends introduced him to a local broker who claimed to have experience in sending labourers abroad. The local broker convinced Ali to seek overseas employment. Later, the broker convinced Ali's father as well. Ali decided to move after his father gave his consent. They sold their land to finance his migration. Ali paid the full amount in a number of instalments.

... I went mad when I sold my piece of land which was our only inherited property on which we counted [on] for food grain. I was afraid I could never get back the land if I missed it. I became desperate to go abroad in any country. I was kept waiting for three months after I paid the full amount of money. Obviously if I couldn't go abroad I couldn't buy back my land. No one wants to lose ancestral assets; however, it was me who did.

This section reports on the amount of money the migrant workers borrowed to cover the cost of migration. My data show that a majority of both the BRHKs and the BRMs borrowed money to support their migration, although there was a significant difference in the amount: The mean amount borrowed by the BRHKs was Tk. 31,903.79 (USD 469), while the mean amount borrowed by the BRMs was Tk. 49,285.71 (USD 725). This indicates that the BRMs had a significantly higher amount (P<0.004) of debt to external sources compared to the BRHKs. The socio-economic background of the BRMs may have played a role in this disparity since this population has a relatively lower economic background, as compared to the BRHKs. A quarter of both sets of respondents borrowed additional money to support their families left behind until they could start remitting money (there is usually a period of time between the migrant's arrival in the host country and the receipt of income earned), which was not included in the calculations for this study because most of the respondents reported that they did not remember how much was borrowed. This system of borrowing money from diverse sources has caused both a financial and a psychological burden for the migrant workers. According to respondents' reports, the pressure to pay off the loans, compounded by fear of losing the collateral disturbs their mental peace. Thus, judging from the average debt owed by each of the study populations, we can anticipate that the BRMs pay higher psychological costs than the BRHKs.

This section shows how many borrowers (BRHKs and the BRMs) have been able so far to pay off their loans and the average amount of time they needed to pay them off.

My data indicates that the majority of the migrant workers did repay their loans, although across varied durations of time. An overwhelming majority of the BRHKs reported paying back their loans to lenders, while only 69 percent of the BRMs had done so at the point of the interview. However, those who had not yet paid back their loans were relatively newer migrants i.e., most of the loan defaulters had been in the host country for a relatively shorter period of time compared to those who had been able to repay their debt punctually. The notably higher percentage of BRHKs who reported paying off their loans indicates that the BRHKs had a higher financial capacity, and/or lower debt than the BRMs.

... salary that I receive is half the promised (RM475). I borrowed Tk. 55,000 (USD809) from friends and from a money-lender. In two years time I paid back only Tk. 32,000 (USD471). I don't know when I can pay back the rest. This gives me a lot of mental stress. (Alauddin, migrant worker in Kelang, Malaysia)

The repayment status shows that many of the study respondents did not have sufficient income to repay their loans. This indicates that the inability to pay down their debt is not only due to the fact that they were in Malaysia for a shorter period of time, but also because they were under-paid. In addition, the exorbitant migration cost contributed to the default status.

Table 5.4 Amounts borrowed (in Taka)

Amount	Hong Kong		Malaysia	
	n=32	%	n=41	%
10000–20000 (US$147–294)	13	40.63	13	31.00
20000–40000 (US$294–588)	17	53.12	9	21.95
40000–60000 (US$588–882)	2	6.25	5	12.90
60000–80000 (US$882–1,176)	–	–	11	26.83
80000–100000 (US$1,176–1,471)	–	–	3	7.32
Mean	31903.79		49285.71	
Standard deviation	15848.800		26423.745	
Significance (Hong Kong vs Malaysia)	P<0.004			
Total	**32**	**100.0**	**41**	**100.0**

Note: '–' indicates not applicable.
Source: Author's field data, 2004–2006.

As already mentioned, borrowing money has long term effects on all aspects of life for migrants and their families left behind. The duration of time necessary to pay back the debts owed is significant to further understanding the impact of the migrants' earnings. Ullah and Routray (2003) explained that the time factor is significant because it specifies the level of income earning by the borrowers, provided that willingness to pay back debt remains unchanged. Borrowers are well aware that the interest compounds with time, therefore, the sooner they can repay the loans, the better it the circumstances will be for them. Therefore, there is a correlation between under-payment and the duration of a migrant's stay – i.e., the more underpaid they are, the longer they need to work in order to pay off their debts, therefore the longer they have to stay away. The following case demonstrates more concerning the issue of time duration. The majority of the BRHKs (55 percent) paid their loan back within one year of their migration, while 19 percent paid back their debt within three years. Only a very few of the BRMs (5 percent) took less than six months to pay back their loan, while nearly half (45 percent) paid everything back in one year, and the rest paid their debt off within three years of their departure.

Abul has been in Malaysia for the last 13 years and had spent a total of Tk. 120,000 (USD 1,765) for his migration. He borrowed Tk. 40,000 (USD 588) from one of his relatives and the remaining Tk. 25,000 (USD 368) was borrowed from a money-lender. He paid back the money in three years through a number of instalments. With the interest, the amount rose to Tk. 32,000 from Tk. 25,000. However, Abul is still not able to pay the money back which he borrowed from his relatives. Abul reports that he currently he has no savings. After he lost his full-time job at the paper mill in Selangor, Malaysia he moved to the Shah Alam area where he found many compatriots. He managed to obtain a part time job, although he was not able to find a full-time job again. He often manages to find additional part-time work on an hourly basis in different shoe and paper factories. While he made several attempts to return home, he failed to gather the money required to purchase an airplane ticket. This example of delayed repayment can be explained by the higher rate of interest on the principal loan and the loss of a full-time job in the host country.

The psychosocial cost and rationalizations of migration decisions

While migrant workers make a financial difference to the family members left behind (as will be discussed in the next chapter) – sometimes the difference between basic survival and a better life – there is a social cost involved. While it is argued that migrant workers often inflict a number of social costs on the receiving society, they, themselves, pay high social costs as well. For example, migrant workers' children are raised in single-parent homes and, since culturally the discipline role usually belongs to the father, and males are more often those who migrate, this can cause difficulties and may result in behavioural problems or in children falling into bad company (Toms 2004). These psychological and social costs impact on the migrants' process of rationalizing their migration decision. Those who have migrated often think later that the money they earned cannot compensate for the psychological cost they pay personally while abroad, and the social cost they and their families might pay.

There are many aspects associated with migration which are difficult to measure in quantitative terms (IOM 2005). The non-economic costs – social and psychological – are such examples. While the monetary costs are limited to out-of-pocket expenses, the non-monetary costs extend far beyond the migratory experience itself, and may continue for years after return, including the psychological cost of relocation, as well as being treated as a marginalized member of society in the destination country. The monetary benefit consists of the real earnings achieved through migration. However, the cost and benefit must be analyzed from personal and social perspectives as well as purely economic ones. Personal costs are those affecting an individual or a family, while the social costs affect society as a whole. No equation, to my knowledge, so far has evolved to measure these factors in quantitative terms (Miyan 2003). I have developed a set of psychological variables, which, in this study, have been described in terms of frequency and compared between the BRHKs and the BRMs in the following Table. A migration decision that involves a choice between options 'to stay' and 'not to stay' has an opportunity cost.[9] Due to leaving their place of origin migrants have to forego other alternatives that would have been available to them had they stayed. Furthermore, once migrants realize the actual cost of a chosen alternative, they may begin to contemplate the true costs of their migration, both monetary and non-monetary, which may increase their sense of psychosocial stress. Every migrant interviewed for this study felt that he/she has had to pay a psychological price while abroad, due to estrangement from their kin group at home, among other factors.

My data show that the BRMs suffered more as compared to the BRHKs by all the psychological indicators used. A significantly higher percentage of the BRMs suffered anxiety than the BRHKs ($P<0.001$). This indicates that the ability of the BRHKs to visit their families more frequently caused them to suffer less stress. This was impossible for the BRMs due to the lack of money, visa related complications, high travel expenses and a lack of leave days. Besides contact with family, the migrants' work load, relationship between employers and employees, wage levels and legal status are often linked with felt stress. While all indicators of depression were endorsed by some proportion of the study respondents, the three most pronounced problems were anxiety, depression and, for BRMs, changed sleeping patterns.

The preceding sections have provided both qualitative and quantitative analysis about the making of migration decisions: The fact that expectations of economic gains and a better future prompted these workers to make their decisions to migrate and work overseas but that these hopes have remained, in most cases, unrealized. This section analyzes the current views of respondents as they try to justify their migration decisions by looking at migration as a way of enhancing their prestige in their community at home. This has been further analyzed as to whether they think their migration decision was the best available option, including what were their forecasts about migration, and what were the factors that influenced them to make their migration decisions.

This section examines the rationalization of migration decisions in terms of prestige. Although a relative term, prestige is one of the basic psychological needs of human beings, and contributes to feeling dignified in a given society. A lack of prestige impacts many aspects of people's lives because it is closely related to the economic level of an individual or a family in a society. As people make decisions and act, they do so in an effort to acquire a certain sense of prestige (Branden

---

9   While the cost of a service is often thought of in monetary terms, the opportunity cost of a decision is based on what must be given up (the next best alternative) as a result of the decision. The advantage forgone as the result of the acceptance of an alternative should also be considered. Whatever course of action is chosen 'to stay' or 'to move' the value of the next best foregone alternative course of action is considered the opportunity cost.

1997). Several studies have examined the issues of prestige and argued that prestige must be considered a significant factor in the rationalization of migration decisions (Rahman 2003).

Table 5.5 The psychological costs (multiple response)

Psychic indicators	Hong Kong		Malaysia		Significance (HK vs M)
	f	% (n=56)	f	% (n=70)	
Tension/stress/anxiety	12	21.43	53	75.71	P<0.001
Physical illness	18	32.14	41	58.57	P<0.003
Increased level of depression	29	51.79	64	91.43	P<0.004
Changes in sleeping pattern	8	14.29	51	72.86	P=0.000
Changed motivation	11	19.64	48	68.57	P<0.002
Emotional change	15	26.79	34	48.57	P<0.002
Decreased social activities	3	5.35	26	37.14	P=0.000
Increased anxieties for those left behind	12	21.43	29	41.43	P<0.001
Tension about education of the children	19	33.93	33	47.14	P<0.003
Tension about land cultivation	13	23.21	22	31.43	P<0.003
Feelings of low-esteem	11	19.64	43	64.42	P=0.000

Source: Analysis results from author's field data, 2004–2006.

It is important to note that an overwhelming majority of the BRHKs (93 percent) considered migration to have enhanced their prestige in the locality, while only around a third of the BRMs had the same opinion. Many respondents stated that money was a critical factor in enhancing their prestige, so they felt that if they could send money home and be seen to spend more liberally, their prestige would be further enhanced. Carling (2005: 18) argues that the social position of migrant families who receive remittances does much to galvanize the family economy and adds value to the family's sense of being respected. However, a different view becomes evident in the current study. Many of the respondents thought that their neighbours and other local people took on the perspective that they, as migrant workers, were not fit for survival in their own locality and that this was the reason they moved overseas as labourers. A few of the respondents expressed the opinion that the prestige factor was significant a decade ago, but not now:

> Things have changed. Villagers used to come to see a returnee ... This no longer
> happens. People nowadays understand that the "unfit" move to work in foreign
> countries, hence, to them, it is no longer a matter of prestige ... (Nazrul, a
> migrant worker, Seri Serdang, Malaysia)

Like Nazrul, many other BRMs did not consider staying in Malaysia prestigious. This demonstrates a significant difference in the perception of migration if prestige plays a more major role for the BRHKs than the BRMs: BRMs observed that there was not much visible impact of migration on the overall family economy. Therefore, for them, the prestige issue was not significant. This demonstrates a clear difference in opinion on the importance of prestige between the BRHKs and BRMs: A significantly higher percentage of the BRHKs thought that it had enhanced their prestige in their locality. However, it appears from this study that this is because the BRHKs had been able to contribute to their family in a number of ways, by enhanced living standards, increased consumption expenses and renovated houses, and therefore the impact was visible to the community.

According to the value expectancy perspective, individuals as rational actors choose the option among alternatives that they believe is likely to give them the greatest satisfaction (Heath 1976, Carling 1992, Coleman 1973, Roemer 1988, Elster 2000). Therefore, this section deals with a question of whether there were alternative choices for the BRHKs and the BRMs other than migrating to HK or Malaysia. Highly similar percentages of the BRHKs and the BRMs (32 percent of BRHKs; 27 percent of BRMs) reported having alternative choices prior to migration: among both sets of respondents, it is clear that a majority felt that there were no alternative choices at the time of making their migration decision. Respondents considered migration to the destinations they chose to be the best alternative, despite the problems and hardships they have encountered. BRMs chose Malaysia by taking a number of factors into consideration, such as ease of adapting to the society, and kinship and friendship ties, as well as economic considerations. BRHKs, however, chose HK because, to them, it offered better economic returns, and they could stay in touch with family and return home easily.

Respondents reported going through a process of making forecasts, asserting their expectations and considering some of the possible consequences of migration before making the journey overseas. This section examines the forecasts that the study respondents made about the country of destination, job and salary. In addition, respondents were asked if they had forecast post-migration consequences, and if so, what those forecasts were, and whether they had influenced their migration decisions. Further analyses of these forecasts identified the gaps between pre-migration expectations and post-migration realities.

The interviews revealed a number of common forecasts that respondents had made before migrating. Respondents' forecasts fell into four main categories, shown in the following Table. Building a bright future was the most pronounced forecast of both the BRHKs and the BRMs. A few (10 percent) of the BRMs said that they thought they could catch the so-called 'golden deer' if they could move out of the country, which was a response not endorsed by the BRHKs, although evidence in the media indicates that this assumption is shared by many other migrants (*Daily Ittefaq* 2007). Many of the BRHKs (29.39 percent) were motivated by the possibility of further migrating to other countries, especially in Europe, using Hong Kong as a gateway, which was not the case with the BRMs.

Several respondents expected to have to take up illegal work as they did not have a set job offer prior to migration, although it is interesting to note that significantly more BRHKs (30.36 percent) had considered this possibility prior to migration than BRMs (11.43 percent).

Further analysis has been done to identify gaps between these forecasts and post-migration realities. Figure 5.2 shows how many of the respondents found their forecasts to have been accurate post-migration. Data show that 19 of the BRHKs and the 38 of the BRMs had found their forecasts to be incorrect, while 30 and 32 of the BRHKs and BRMs, respectively, felt that their forecasts had materialized. Most of the BRMs said they realized that their forecasts were incorrect only after they had made the journey to their destinations, when it was already too late for them to revise their decisions. This indicates that they did not have accurate information during their decision-making process and before they left Bangladesh. This coincides with the data from Chapter 4 concerning the use of illegitimate brokers in order to facilitate migration, and, in fact, many BRMs felt that they had been trapped and deceived by brokers who took advantage of their unemployment, illiteracy, and limited access to information.

Who makes migration decisions? Although in many cases, the decision to migrate happens after discussion and negotiation with the family members, according to the study respondents, actual migration decisions were not dependent on familial consent. Even in cases where family members have given their consent, they frequently are no better informed than the migrant themselves, and do not know whether the decision will be a good one or not. Some family members gave their consent only because they could not forbid or stop a family member from going: There are certainly cases in which some family members did not agree to the decision, but the migrant was adamant, rendering the participation of family members in decision-making void, if not useless.

One intriguing corollary to the high proportion of economic migration in Bangladesh has been the emergence of female-headed households and the change in traditional patriarchal attitudes towards gender roles and women's status. This has created an opportunity for greater involvement of females in household decision-making. However, participation of females in household decision-making is in prospect when women assume control over the household resources. The prolonged periods of absence of male household members is likely to be a significant factor behind this. Thus we can observe a potential cyclical pattern in which the absence of men due to migration, enhances women's roles in household decision-making, which therefore gives women a greater say on future migration decisions. Therefore, it is not as easy to shift through the decision making process of migrants; whether the decision and the action to migrate is a collective call or just an individual preference. This is likely to vary from household to household and to depend on a number of personal and familial factors.

The section analyzes the opinions of migrant workers concerning their migration decisions. Based on the wisdom of hindsight, the interviewees were asked directly whether they thought that their decision to migrate had been correct?

In response, an overwhelming majority of the BRMs (69 percent) expressed that their migration decision had been incorrect and only around 27 percent thought it had been correct. In contrast, nearly half of the BRHKs (45 percent) thought their decision to migrate to Hong Kong had been correct and 43 percent considered their decision to have been incorrect.

When asked why they felt their decision was correct, respondents put forward numerous explanations to justify their migration decision: one-third of the BRMs expressed that at least they had been able to have some form of employment, regardless of income. Others rationalized their decision by stating that staying in Bangladesh was not 'safe' anyway, but of these, most declined to explain what they meant by 'safe'. Those who did provide an explanation stated that there were threats to their lives in the home country. Most of the BRHKs provided similar explanations to the BRMs; unemployment at home was regarded as the most significant justification to them.

Perceived forces affecting migration decisions

This section ranks the main perceived forces affecting migration decisions by applying a priority index. Responses both from the BRHKs and the BRMs were merged together as the variables emerged from both the groups did not differ significantly. The variable 'search for work' ranked first with the highest Weighted Mean Index (WMI) (0.753) followed by the 'marital factor' (WMI = 0.738). 'Relative deprivation' ranked third with a WMI of 0.729, followed by 'higher income probability' with 0.726 WMI. 'Brokers' inducement' ranked 5th with 0.716 WMI, followed by 'chances to obtain ingress into European countries either illegally or legally (WMI = 0.70). Interestingly, most migrants appeared to believe that once they entered a European country, they could stay long term without any problem (*Daily Ittefaq* 2007). 'Threat from political forces' received 20th rank with 0.375 WMI, while 'relatives' inducement' ranked the last with the lowest WMI (0.339). The respondents have prioritized 'search of work' as the number one factor that induced their migration decision. This resonates with the findings in Part III concerning the previous occupational status of the migrants (predominantly unemployed). However, the marital factor has become an important one to consider among migration predictors: Migrants shoulder extra familial burden due to being married, therefore, to respondents, marital status plays a role in inducing them to migrate. All variables with corresponding WMI include: Search of work WMI = 0.753; Marital factor (married) WMI = 0.738; Relative deprivation WMI = 0.729; Higher income WMI = 0.726; Induced by the brokers WMI = 0.716; To sneak into European countries WMI = 0.70; Information on foreign countries WMI = 0.693; Nothing to do at home country WMI = 0.687; Marital factor (unmarried) WMI = 0.679; Inadequate information WMI = 0.665; No education/no training WMI = 0.649; To use the savings WMI = 0.633; To settle a disputed piece of land (sale) WMI = 0.619; Reluctance to work in rural areas WMI = 0.618; Loss in previous business WMI = 0.602; Pressure from family WMI = 0.549; Escape conviction WMI = 0.534; Joining relatives WMI = 0.485; Desire to settle abroad WMI = 0.517; Threatened by political forces WMI = 0.375; Induced by relatives staying abroad WMI = 0.339.

The factor analysis of migration predictors

This section tries to group the many predictors determined by the priority index above into a few broad categories. The application of factor analysis has been found to be the best tool for this purpose. The factors that trigger migration decisions, as perceived by the study respondents, were extracted through this analysis: The higher the commonalities of the corresponding variables, the stronger are the determinants of migration decisions. A scree plot[10] was used to determine the number of migration predictors, and reducing them to three key factors: economic, socio-political, and psychological. Economic factors cluster around those forces related to financial status, while the variables that tend to explain socio-economic issues were grouped into the socio-political factor. Economic factors explain 53.255 percent of the total variance (Eigenvalue is 2.663), indicating the strong explanatory power of this factor, followed by socio-political factors explaining 46.942 percent of the total variance (Eigenvalue is 2.816). The psychological factor has been the most powerful in explaining the rationalization of migration decisions: This factor explains 61.888 percent of the total variance with an Eigenvalue of 4.330.

Multiple regression of migration decisions

Variables influencing migration decisions of the respondents have been analyzed by applying multiple regressions. This analysis has been done to determine the predicting variables that triggered migration decisions in order to determine whether there is convergence between the study data and predictors assumed in the theoretical frames. Here, the dependent variable is the migration decision. The independent variables are:

---

10   This is performed (in factor analysis technique) as a data reduction device to determine the 'number of factors' depending on the diversity and volume of the variables.

p_1	Expectation of higher income	p_{12}	Desire to settle abroad
p_2	Search for work	p_{13}	Desire to sneak into European countries
p_3	Pressure from family	p_{14}	Desire to settle a disputed piece of
p_4	Induced by brokers		land (sale)
p_5	Induced by relatives to go abroad	p_{15}	Ability to use savings
p_6	Joining relatives	p_{16}	Unemployed in home country
p_7	Information about foreign countries	p_{17}	No education/no training
p_8	Relative deprivation	p_{18}	Marital factor (unmarried)
p_9	Threatened by political forces	p_{19}	Marital factor (married)
p_{10}	Escape conviction	p_{20}	Loss in previous business
p_{11}	Inadequate information	p_{21}	Reluctance to work in rural areas

These above variables were considered independent variables, and were based on the perceptions of the respondents. Correlation analysis was performed for variables demonstrating a high correlation with dependent variables. For maximizing the relationship between the dependent and independent variables, the variables with a correlation coefficient smaller than 0.3 or $R^2 < 0.09$ were removed from the model. The remaining variables are shown, along with their correlation coefficients hereunder.

Correlation Coefficients (analysis results from field data, 2004–2006)

Variables	Correlation Coefficient
Expectation of higher income	0.415
Relative deprivation	0.345
Unemployed at home country	0.340
Information about foreign countries	0.315
Search for work	0.310

The variables with a correlation coefficient of more than 0.3 were entered into a stepwise multiple regression analysis to minimize the number of independent variables and maximize the level of explanation. The regression coefficient appears below which constructs the regression equation. Five variables are included in the model. All of the variables are significant at 95% confidence level.

$Y = 7.9803 + .289* p_1 + .310* p_2 + 1.543* p_8 + 2.260* p_{16} + .392* p_7$ (Only p_1; p_2; p_8; p_{16}; p_7 were found to be significant predictors)

The R^2 value of 0.734 indicates that this model is useful in analyzing the factors which influence migration decisions that are determined by the five variables elicited by the factor analysis: expectation of higher income, search for work, relative deprivation, unemployment in home country, and information about foreign countries (i.e. networking). The findings show that these variables explain migration decisions effectively (all have p-values under 0.02). This establishes the fact that the issues discussed in the previous chapters are strong enough to explain the rationalization of migrant's decisions. This analysis also endorses the findings extracted in the factor analysis. More importantly, this analysis has validated components of the theoretical framework proposed in this study by extracting the analogous variables for triggering migration decisions.

Table 5.6 Regression coefficients

Variables	Coefficients	T Value	Significance
Constant	7.9803		P=0.000
Unemployment at home country	2.260	4.835	P=0.000
Relative deprivation	1.543	4.230	P<0.000
Information on foreign countries	.392	4.012	P<0.001
Search for work	.310	3.900	P<0.001
Expectation of higher income	.289	2.670	P<0.002

Source: Analysis results from author's field data, 2004–2006.

Promises and realities

Promises are manifestations of intent to act that will or will not occur. In this case, promises are written or oral declarations given to a promisee (i.e. migrant) in exchange for something of value that binds the promisor (i.e. recruiting agents) to facilitate the migration (Dictionary of the English Language 2004). The many pre-migration promises made to the respondents by recruiting agencies or brokers built up their expectations and contributed to their decision-making process to migrate (*Daily Ittefaq* 2007). Mainly two kinds of promises prompted the decision to migrate: a secure job and a 'good' salary. Data, however, show that the majority of both the BRHKs and the BRMs did not get the jobs they were promised or those that they expected. Both groups blamed the misleading information given to them by the brokers for much of the stress resulting from this fact. They stated that they should not have depended so much on the brokers and recruiting agents to facilitate their migration. However, the BRHKs care less about the false promises than the BRMs and are not concerned with these or their built-up expectations anymore. This is likely due to the relatively higher financial return of their migration decision, as compared to the BRMs, evidenced by the fact that, although only 11.54 percent received the job that they were promised prior to migration, 50 percent report receiving the salary they were promised (as opposed to 21.42 percent of the BRMs.

Table 5.7 Promises of job and salary before the move

Job*	Hong Kong		Malaysia	
	n=56	**%**	**n=70**	**%**
Yes	6	11.54	15	21.43
No	46	88.46	49	70.0
NR	–	–	6	8.57
Total	*52*	*100.0*	*70*	*100.0*
Salary**				
Yes	28	50.0	15	21.42
No	28	50.0	47	67.15
NR	–	–	8	11.43
Total	*56*	*100.0*	*70*	*100.00*

Note: '–' indicates not applicable; *Question: Did you get the same salary as promised before moving?; **Question: Did you get the same job as promised before moving?
Source: Author's field data, 2004–2006.

As discussed in the preceding chapters, some of the migrants had relied on syndicate networks for their migration. Most of these were enticed by false promises of a better job. When asked what they believed to be the reasons for the false promises made by the recruiting agents diverse and interesting information emerged (following table). It is clear that there is no reliable recourse for migrants in cases of misinformation and false promises made purposefully by recruiters: data show that the highest percentage of the BRHKs (60.71 percent) report that the dishonest recruiters were never punished by law, and nearly half of them said that the fraudsters escaped easily if even they were caught. The highest percentage of the BRMs (61.43 percent) thought that the brokers or recruiters continued to cheat people because they do not have to fear any legal recourse, while a majority (56 percent) also felt that fraudulent and deceptive methods had become normal practice, and hence were regarded as trivial.

This section seeks to explore the intention of the migrants as to the preferred length of their stay in the host country. Intention of stay a longer or shorter period is related to how the migrants justify their migration. Therefore, respondents were asked how much longer they intended to stay abroad. Theoretically, the behavioural implications suggest that migrants may compensate for lower wages in the host country by prolonging their migration duration. Therefore a decrease in expected wages, to the extent that it is not foreseen, should lead to a compensating change in migration durations (Law 2002). In answer to the question of intended duration of stay, the highest number of BRHKs (78.57 percent) and approximately one fourth of the BRMs (24.3 percent) expressed their intention to stay abroad as long as work is available to them, further indicating the significance of economic factors in migration decision-making. The BRHKs were significantly more precise and decisive in stating that they would stay as long as employment was available than were the BRMs, who express much more ambiguity about their situation. A full half of the BRMs (50.0 percent) stated, usually with disappointment, that they did not know what to do in future: They were unsure about their intended tenure in Malaysia, evidence of their disillusionment with their current situation. Only a handful (14 percent) intended to stay long-term. This data further indicates that the BRHKs consider the work opportunity in Hong Kong to be a significant attractant factor in prolonging their stay while this is not the case for the BRMs in Malaysia's often-restrictive job market.

Table 5.8 Perceived reasons for false promises* (multiple response)

Perceived reasons	Hong Kong		Malaysia	
	n=56	%	n=70	%
Fraudulence and deception**	17	30.37	39	55.70
Employer took advantage**	5	8.93	28	40.0
Agent's deception + employer's whim	6	10.71	24	34.29
Wrong/misinformation	22	39.29	32	47.14
We were illiterate	4	7.14	29	41.43
No legal base of the brokers	13	23.21	43	61.43
Fraud escapes easily	23	41.07	19	27.14
We cannot seek redress	21	37.5	22	31.43
No exemplary punishment given to the fraud	34	60.71	17	24.29
No idea	19	33.93	21	30.0

Note: * Question: What do you think are the reasons for deceiving with promises?; ** Deception by the brokers and the agents.
Source: Analysis results from author's field data, 2004–2006.

By applying the WMI, the level of satisfaction for 'living overseas' has been worked out against the indicators, such as 'work environment and load', 'living place', and 'employer-employee relation', revealing a contrast scenario between the BRHKs and the BRMs. The BRHKs ranked their top source of satisfaction as 'living in Hong Kong', followed by the 'work load and environment', and 'employer-employee relation'. However, living expenses received a negative (-) WMI indicating that they are not at all satisfied with the living expense in Hong Kong. In Malaysia, 'living expenses' ranked top as highly satisfactory, followed by 'living here' in general. However, 'work load and environment' and 'employer-employee relation' received negative (-) values, which imply that the respondents were not at all satisfied with these indicators. This information is concordant with the respondents' reports of various forms of abuses and exploitations in the workplace.

Table 5.9 The level of satisfaction over living abroad

Variables	Highly satisfied W=+2.0	Satisfied W=+1.0	Neutral W=0	Not satisfied W=-1.0	Highly dissatisfied W=-2.0	Σf	WMI	Rank
	(f_1)	(f_2)	(f_3)	(f_4)	(f_5)			
Hong Kong								
Living here	20	2	4	8	4	38	0.684	1
Work load/ environment	22	13	4	12	5	56	0.625	2
Employer/employee relation	21	3	7	8	7	46	0.5	3
Living expenses	11	2	23	14	6	56	-0.036	4
Living place	9	7	13	9	12	50	-0.16	5
Malaysia								
Living expenses	18	12	3	10	1	44	0.818	1
Living here	13	9	12	11	8	53	0.151	2
Living place	5	13	10	4	7	38	0.0263	3
Work load/ environment	9	2	19	15	20	65	-0.538	4
Employer/employee relation	7	1	8	29	14	59	-0.711	5

Source: Analysis results from author's field data, 2004–2006.

PART III
Rationalizing Post-Migration
Decisions

PART III

Chapter 6
Working and Living Conditions in the Host Countries

The key objective of this chapter is to analyze the significance of the living and working conditions of migrant workers as they relate to the rationalization of post-migration decisions. In the preceding chapter, I addressed the pre-migration issue of financing and we now turn to living and working issues of the initial phase of the post-migration process, which migrant workers face immediately after arriving at their destination countries. In these new environments, migrant workers encounter innumerable challenges in living and working conditions which may often be considered as exploitative by every measure of human rights standards.

This chapter is divided into three sections. The first deals with diverse issues of employment such as categories of work, wages, working conditions and employer–employee relations. The second section deals with the experience of living in host countries, regarding a migrant's length of residence, legal status and comfort factors in his or her places of habitation. In the third section, the focus shifts to different strategies of adaptation and assimilation migrants utilize in their host-countries, and analyzes the factors that influence migrants' post-migration decisions.

A number of studies have taken living and working conditions as two significant factors for making post-migration decisions (IOM 2005). Migrants' intentions to prolong their stays in destination countries depend largely on the extent and process by which the workers may adapt and assimilate culturally, economically and socially. The assimilation of migrants into the Malaysian and Hong Kong labour markets, for example, is influenced by various factors, such as the migrants' ability to obtain and maintain work permits, the type of work they normally manage to acquire, the neighbourhoods they work in, and finally, the state of their living conditions.

While looking at their working conditions and wages, we take a deeper look at the economy because migrants face a disadvantage in initial earnings relative to their native-born counterparts (Zeng 2004). These are exacerbated by certain barriers to assimilation such as language and ethnicity until migrants can adapt within their destination countries.[1] Assimilation is a process of integration whereby

---

1   Assimilation theory is based on neo-classical economics and has historical roots in the urban experience of earlier migrant waves. The assumption of the model is that new migrants are young, with limited resources, and cluster together in low-income migrant enclaves (Logan, Alba and Zhang 2002) for both economic and social reasons (Myles

migrants are absorbed into a host community (Bohning and Beiji 1995, Dorall 1989). Dustmann (1993) observes that both permanent and temporary migrants, however, overcome this disadvantage only after they adapt to the receiving society. The flaw of Assimilation Theory is that it applies only to migrants who can become residents and citizens in their new countries. Therefore, this theory is not for transient labour migrants who are not allowed to stay and who, moreover, often become illegal migrants. Despite the fact that illegal migrants cannot be fully assimilated due to the structural barriers to immigration status, they can nevertheless adapt to local conditions in terms of coping strategies that enable them to maximize their chances in the host countries. Adaptation is therefore more applicable to illegal migrants than is assimilation.

Employment and working conditions

In this section, I analyze a range of issues related to employment including migrant workers' contracts, the categories of jobs available to them, their salaries, the challenges they face and other aspects of their employment.

A valid work contract guarantees the migrant workers' rights to life insurance, medical care, rest days, repatriation after the contract expires and treatment in compliance with conventions of human rights (see ILO 2004). The contract thus benefits risk groups by encouraging labour inspection to enforce minimum decent work condition. In many cases, however, recruiters do not bother with a contract. Some employers prefer to recruit workers without making any formal agreement (i.e., illegally) circumventing regulations, and depressing wages. In some countries, such as Malaysia, there is no legislation on minimum wage (Abdul-Aziz 2001, Piper 2005, Jamil and Rebecca 1993). The contract of overseas work is negotiated when the formal processes of recruitment are followed and two-year contracts are the norm for most of the migrant workers (INSTRAW/IOM 2000).

This research shows that an overwhelming majority (79 percent) of the migrants (BRHKs) had no direct contract with their employers. There was no point signing an agreement or contract as they worked on an hourly basis and met new employers everyday. The BRHKs seek work on street-corners and employers take them to their places of work as and where needs arise.[2] Consequently, migrants

and Feng 2004). Regions where such assimilation occurs have often been referred to as a 'melting pots' (Southwest Economy 2004). Many studies however connect adaptation and assimilation with many aspects of working and living places (Valdez 1999). It is notable that this theory assumes that migrants can become permanent residents and citizens, however, which is not applicable to transient labour migrants.

2   The employers slow down their cars a few feet away from the spots where the workers wait. Workers rush to the window of the car. They don't waste time negotiating the wage rate while on the street: They often do that while on the car or at the work place, sometimes on phone beforehand (if known) to avoid police gaze. Often the employers signal from a little distance to approach them. Employers who hire the migrant workers generally

work with one employer for one or two hours with the wage on hourly paid basis. Only a few of the workers employed by hotels, tailoring shops and watch shops as agents had contracts. This means that possession of a contract or agreement is dependent on the type of work offered or available to the BRHKs. Data show a significant relation between the category of work and the possession of a contract (P<0.013). In contrast, the BRMs (except those having contract) go to factories to look for employment. Data show that majority of the BRMs worked as daily labourers in different factories without any contracts.

As already discussed, there has never been an agreement for labour export from Bangladesh to Hong Kong. The responsibility of signing a contract, then, lies solely with individuals who obtain a job in Hong Kong. In contrast, there exists a labour export agreement with Malaysia although it remained ineffective from 1999 to 2006. Until the point of time (including the period of the present interviews in 2006) – the sedentary existence of agreement of labour export in Malaysia and following the clean up drive[3] to reduce irregular migration in 2005 – Bangladeshis were not allowed to enter the Malaysian labour market. Migrants were not oblivious to this fact: during this period of time, a group discussion revealed that the workers were aware of the need for valid documents in order to help mitigate their vulnerabilities.

The primary objective of this analysis is to evaluate and present the type of occupation, as well as the work content and load, that migrant workers take up in their host countries. In migration literature, categories of jobs given or available to the migrant workers remain a major issue. With diverse consequences for the migrants and their families left-behind, migration to other countries opens up better opportunities for income and work and access to health and educational facilities. But not all migrants are better off in terms of quality of work; some struggle to find jobs and put up with work that is beneath their qualifications (Ghai 2004) and employment in the 3D[4] category of work (Mardzoeki 2002).

---

know the spots where they gather to seek work. They take them to the work location from the pick-up spots. Knowing it is an illegal practice, the employers hire them clandestinely.

3   In March 2005, the Malaysian government declared general amnesty (repeatedly) for those staying in Malaysia illegally. Many Bangladeshis used this amnesty to accept the 'travel pass' given by the government and managed to repatriate.

4   Dirty, dangerous and demeaning (some researchers say difficult, degrading or demanding). Pécoud and Guchteneire add that migrants' vulnerability is also increased by the sectors of economic activity in which they are active. They often have so-called 'three-D' jobs and are also over-represented in marginally viable and sometimes semi-legal sectors such as seasonal agricultural work, domestic services and the sex industry, in which the protection of workers is underdeveloped (Pécoud and Guchteneire 2004: 3).

Table 6.1 Possession of work contracts by affiliation

Profession	BRHKs Contract			BRMs Contract		
	Yes	No	Total	Yes	No	Total
Daily labourer*	–	37 (71.43)	37 (71.43)	10 (14.29)	47 (67.14)	57 (67.14)
Hotel	2 (1.79)	1 (1.79)	4 (7.14)	–	–	–
Sex workers	–	1 (1.79)	3 (5.36)	–	1 (1.43)	1 (1.43)
Tailoring shop	6 (3.57)	4 (7.14)	8 (14.28)	–	–	–
Fake watch shop	4 (1.79)	1 (1.79)	4 (7.14)	–	–	–
Construction site	–	–	–	–	9 (12.86)	9 (12.86)
Other**	–	–	–	–	3 (4.29)	3 (4.29)
χ^2 Significance		P<0.013				
Total	**12 (21.43)**	**44 (78.57)**	**56 (100.0)**	**10 (14.29)**	**60 (85.71)**	**70 (100.0)**

Note: '–' indicates not applicable; * Plastic factory, shoe factory, garment factory, furniture factory/shop and chocolate factory; ** Printing press; Figures in parentheses indicate percentages.
Source: Author's field data, 2004–2006.

While categorizing the work offered to the BRHKs and the BRMs in light of the 3D model, a picture emerges of migrants occupying the lower echelons of occupations. This reality is partly born of ignorance: Most migrant workers did not have any clue before they moved as to what kind of work they would find abroad, and many could not obtain work immediately. Indeed, most of the BRHKs took around two weeks to find jobs. During this period, they obtained ideas on how to adapt and what type of work awaited them. Most of the BRHKs brought enough money with them to sustain the period until they found employment.

> ... labourers are labourers ... after death we will remain as labourers. We were born to be labourer. (Mansur, a migrant in Malaysia)

This statement says all about the feeling of frustration and deprivation of the BRMs. This also says that those who had low status occupations in country of origin retain their low status occupations in the host society (Mardzoeki 2002). Labour migration from Bangladesh, while providing opportunities for employment, is associated with employment in the low status occupations.

Most job opportunities offered to Bangladeshi workers in Hong Kong were either as contractual labourers or as sales agents. Contractual labour was largely part-time, with payment calculated on an hourly basis and made instantly on the spot. Payment varied depending on the workload and often on the compassion of employers, who were sometimes generous enough to pay an additional reward.

As a contractual worker, one's work covered, but was not limited to: loading and unloading cargo from trucks for wholesale shops and constructions sites; factory work; house cleaning and other domestic chores. A few BRHKs worked as agents of counterfeit watch stores,[5] hotels and tailoring shops. The nature of their jobs included convincing clients – mostly foreign tourists – to buy watches, sealing hotel stay-deals with tourists, and the making and selling of suits. Bangladeshi migrant workers are visible on the streets at Tsim Sha Tsui area with their offers. They are paid commissions, on top of their monthly lump-sum salary, on every deal made with a client.

Photo 6.1 Migrant day-labourer ready to unload lorry

The study shows that 73 percent of the BRHKs are employed as part-time workers from the Sham Shui Po area. It should be noted that the labour classes who work at Sham Shui Po area do not work at Tsim Sha Tsui and Central areas as they are considered too unkempt and dishevelled to work in these areas. This has furthered division in the labour market producing two distinct working groups in Hong Kong: Those who work at Sham Shui Po, who normally get dangerous and dirty work, and those who work at Central, Tsim Sha Tsui and Wanchai areas, who

---

5   Expensive brand-name watches such as Tudor and Rolex are symbols of status and prestige in Hong Kong. Fakes are difficult to distinguish from the real thing and agents are charged with convincing the customers to buy the fakes.

get difficult, though less dangerous, work (also see Achariya 2003) which, the workers themselves say, 'requires patience'.

Female labour migration from Bangladesh has been periodically restricted since it begun to gain momentum in the late 1970s (INSTRAW/IOM 2000, Surtees 2003). Five of the study's respondents were females who have been working as sex professionals since coming to Hong Kong as tourists. Although the overwhelming majority of the BRMs took up work as labourers, two of the BRMs, who were dependants of other migrants and were later deserted, resorted to sex work for survival.

In addition to the primary job migrants took up secondary jobs during their spare time. Five of the BRMs reported doing secondary work. They purchase tax-free cigarettes on return from China or elsewhere and sell them to peers and compatriots. A few of the BRHKs and the BRMs also carry gold and electronic goods such as cameras and mobile phone sets to Bangladesh for resale. Among other secondary jobs are cleaning factories and selling phone cards to the Bangladeshi community people.

Researchers consider irregular and delayed payment as forms of exploitation (Waldinger 1997, Abdul-Aziz 2001). Therefore, withdrawal of salary and modes of payment issues are significant elements to be taken into account in the current research context. Withdrawal and modes of payment largely vary depending on the nature of employment contracts. Data show that none of the BRHKs reported receiving their salary/wage irregularly. Although they were often underpaid, they reported this as preferable to the situation in their home country. Around one-third of the BRHKs received their bill or salary in cash on a monthly basis and an overwhelming majority received on a daily basis after the job is completed.

Around one-third of the BRMs complained that the employers did not pay their salary in full, frequently a portion of the salary was withheld. Another third claimed that they never received their salary on regular basis. Approximately one-quarter of the BRMs reported receiving cash on a monthly basis. In comparison, all the BRHKs received their salaries in cash after the work was completed, and one-third of the BRMs reported receiving salary through bank transfers. Two distinct characteristics were observed in the modes of withdrawal between Hong Kong and Malaysia. The BRMs preferred their salary to be deposited to the bank in order to avoid it being snatched by the 'Tamils' or by the police. However, it is not trivial for BRMs to open an account at, or to withdraw money from, a bank. Many BRMs with bank accounts complained that withdrawing money without a letter of support from their employer was not possible. Most of the BRHKs were against the idea of opening a bank account, the main reason being the uncertainty of their residence and the requirement of a guarantor to open an account.

Unsafe working environments, especially for unskilled labourers, have always remained a challenge. With few exceptions, exploitation, large-scale abuses including torture, sexual abuse of both men and women, dehumanizing treatment and occupational hazards characterize overseas employment (Taran 2000, Taylor

2003, Wickramasekera 1996). In this section, I explore the challenges migrants face in maintaining good working relationships with their employers.

Most of the BRHKs consider the income they earn as the best reward for their work – from 10 to 15 times higher than what they could earn at home. They express pride in a hard-working self-image. Interestingly, many migrant workers without legal status in the host countries admit that they enjoyed working. In Hong Kong at least, this might be in part explained by the higher income earned. While asking them what made them enjoy working there most, many respondents replied with a smile and said 'it is money'. One respondent was just back from a job at Causeway Bay area. He worked there for around three hours and got HK$250. He said:

> ... this is equivalent to that of my one month's salary in the home country. We came here for earning money and doing it, that's it. (Migrant labourer in HK)

Some respondents even say that their employers have a congenial attitude towards them:

> ... when we sweat, they offer coke or beer, when we cannot lift a bale, the employers extend their helping hands, when we ask for an extra on top of the fixed rate, they are generous. (Aslam, a migrant worker in Hong Kong)

They stated that there are also downsides to the kind of work they do, from the long hours they need to wait on streets to get a job offer to the laborious and thankless nature of the work itself.

When asked as to which part of the work they like most, a considerable number of the BRMs candidly said that there wasn't anything to like. However, one-fifth remained silent and a few said that their work was not difficult; one-third were resigned to the idea that at least they have some errands to do rather than remaining unemployed at home.

Many of the BRMs expressed their grievances by describing their Malay as 'inhuman'. When asked about particular experiences that led them to make such statements, BRMs listed forceful confinement in the factory by the employers to avoid either police arrest or bribes for employing illegal workers, dumping many workers into one small living place, mostly in the containers,[6] keeping the salary level unchanged year after year, physical and psychological torture, extreme misbehaviour and verbal abuse. Most serious was forcible confinement in the factories (also see Achariya 2003). They said that due to their 'illegal position' they are low-paid despite performing very well. The grievance is evident from the below excerpt.

> ... the employers benefit from our illegal status but they never increase our wage rate. They are never sympathetic to us. I was promised to increase my salary

---

6   Container for freight forwarding and shipping companies.

based my performance four times, it was never executed. (Alamgir, a worker of
APM factory, Seri Kembangan, Malaysia)

Furthermore, Abdul-Aziz (2001) observed the depressing attitude of the Malaysian
employers toward Bangladeshi workers. He found that, even for serious accidents,
the only course of action taken is very often limited to medical treatment either at
general hospitals or private clinics. Whatever monetary compensation the workers
receive for potential harm or disability on the job depends on the discretion of the
employers.

Employee/employer relations

There is a widespread assumption that the main difficulties migrant workers face
are with their employers. This is because, upon arrival, a migrant worker's main
relationship remains mostly with his or her employer. Therefore, in analyzing the
phenomenon in the study, I look at a few issues such as the relationship between
employers and employees. The employers could be very exploitative, abusive
and dehumanizing, and as a result, the workers often face various injustices such
as wrongful dismissal and salary cuts. Employers have extraordinary powers
to determine salaries, the quality of work environments and working hours.
Davidmann observes that the payment, income and work environment are the
matters that often lead to raging controversy in an organization. Therefore, we
need to deal with matters which determine inequities in the power relationship
between employer and employee (Davidmann 2006, Thompson 1984).

The duration of a migrant's employment with the same employer shows the
level of probity and friendliness of the overall work environment such as contracts,
work load, content, schedule and salary. However, as mentioned earlier, the case
of the BRHKs was different as 41 of the BRHKs had no steady employer. As to the
rest, six have been with their current employer from 4–7 years. Data further show
that around one-third of the BRMs worked in the same factory for less than one
year and majority of them from 1–5 years and a few from 6–10 years with varied
duration of intervals. This variability has been ascribed to the fact that they were
either laid off or fired, or resigned under duress.

A common view is that the rights of migrant workers are related to nationality;
accordingly, aliens historically enjoy little legal and civil protection (Pécoud and
Guchteneire 2004, 2005, WARBE 2001). Law (2002) also observes that attitude
(either exploitative or friendly) often varies depending on the cultural and ethnic
origin of the employers.

This study shows that all the BRHKs received job offers from Chinese–Hong
Kong nationals. Some Indians and Bangladeshi employers offer work occasionally,
but BRHKs feel more comfortable to work with Hong Kong nationals. Many of
them mentioned that payment from Hong Kong employers is normally higher than
that from Indians or Bangladeshis. The majority of BRMs worked for the Chinese

Malaysians, while a few (14 percent) worked for the Malays. The workers felt more comfortable working for Chinese employers than for Indians or Bangladeshis. Although they thought that the Chinese are very demanding, Chinese employers offer better payment than others. This means the level of income earned also depended partly on the national origin of the employers. Therefore it could also be assumed that the level of exploitation and vulnerability is often linked to the national origin factor.

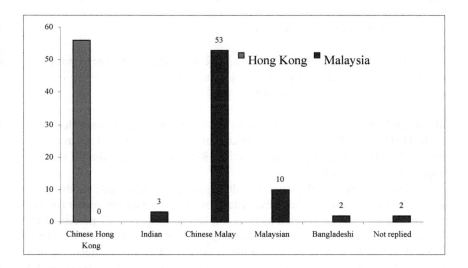

Figure 6.1 Employers by nationalities

This section tries to understand the perception of the respondents on the attitude of their employers toward their workers. Attitude is a psychological response that expresses an individual's positive, negative or neutral views on something (Eagly and Shelly 1993). Pejorative and abusive attitudes of the employers toward the employees are quite common. Abuses of different forms against the migrant workers take place wherever they work, and they occur at every level from recruitment to employment (Jones 1996). There are many examples of migrant workers being forced to repatriate from different countries following various abusive acts. Therefore, the attitude of the employers is one of the significant factors in rationalization. This section deals with one form of the abusive attitude which is 'respect'. The feeling of being respected by their employers is an important factor in assessing the migrants' rationalization of migration decision. The question asked was: 'Do your employers respect you?'

The migrant workers especially the BRMs candidly said that their employers, in most cases, are not friendly. Most of the BRMs felt humiliated and discriminated in many ways for their skin colour and by their class or by the type of work they do. Moreover, in addition to the employers, the Tamils[7] and Malay people insult them by calling 'orang Bangla'.[8] However, a few of the BRMs felt respected by their employers due to their self-assessed qualities, such as being hard working and sincere sense of duty to their jobs. A few of them thought that there was no reason to expect being respected by their employers. One migrant replied that, 'No one respects the poor people and it is universal'.

As Rahman, a migrant worker, said:

> Nobody respects us because we are poor, if we were rich, we would never come to sell our labour at cheap price, I would like to die in the country among our beloved ones, not abroad like a dog.

In contrast to the BRMs perception of the level of respect they commanded, most of the BRHKs felt that they were respected and not discriminated against. Data further show that a higher percentage of the BRHKs compared to the BRMs were respected by their employers.

This section presents the answer to the question: Is the attitude conducive to stay for extended period? The above data show that the majority of the BRHKs thought that they were respected by their employers, while only 27 percent of the BRMs felt respected. Data were further analyzed to demonstrate if this esteem factor is conducive to their staying for longer period, although the esteem factor is not the sole determinant of their intention to prolong their stay. Data also show that their perception of esteem has no impact on their preference to stay in Hong Kong. While the esteem factor is seen as an important one, the income factor is influential in deciding to stay in Hong Kong. Almost a similar situation applies to the BRMs. Furthermore, a higher percentage of the BRMs thought that the esteem factor was not an important variant to facilitate a decision to stay in Malaysia.

---

7   Research on multi-racial nations, such as Malaysia, tends to describe the interracial situation in broad generalizations concerning the major divisions of the population, using such large categories as 'Malay', 'Chinese' and 'Indian'. However, as is well known, trading contacts between Tamil Nadu and the Malayan peninsula extend back over many centuries, the period most relevant here began with the establishment of British control over Malaysia, and the migration of Tamils from Tamil Nadu and Ceylon. Several thousand Ceylon Tamils came to Western Malaysia during the last quarter of the nineteenth century and the first quarter of the twentieth, with a very large portion of the employed male immigrants entering government service at the middle-management levels (Glick 1968).

8   Bangladeshi labour migrants, locally known as 'Orang Bangla' continue to serve Malaysia to perpetuate its status as an 'economic tiger' in Asia. 'Orang Bangla' is generally used albeit in a derogatory sense (Islam et al. 2006).

Table 6.2 Preferences to stay by esteem factor (cross table)

Preferences*	Esteem factor				Total
BRHKs	**Yes**	**No**	**Maybe**	**NR****	
Yes	9 (16.07)	9 (16.07)	9 (16.07)	–	
Not good/not bad/so so	16 (28.57)	6 (10.71)	2 (3.57)	5 (5.36)	56 (100.0)
Total	*25 (44.64)*	*15 (67.86)*	*11 (119.64)*	*5 (5.36)*	
BRMs					
Yes	6 (8.57)	10 (14.29)	4 (5.71)	3 (4.29)	
Not good/not bad/so so	13 (18.57)	19 (27.14)	–	14 (20.0)	70 (100.0)
Total	*19 (27.14)*	*29 (41.43)*	*5 (7.14)*	*17 (24.29)*	

Note: '–' indicates not applicable; * Question: Do you like living here?; ** No response;
Figures in parentheses indicate percentages.
Source: Author's field data, 2004–2006.

The work-day loss

Four factors have been identified as causally connected to work-day losses, which have become inescapable in the overseas life of the migrants. (However, the first factor 'days spent in PR China' does not apply to the BRMs.) The other three factors: Days spent in home country; in jail; and unemployed have been analyzed for both the BRHKs and the BRMs.

Data show that the BRHKs were unemployed for varying number of days and for varied reasons. Work-day losses caused due to illness could not be calculated because, in most cases, respondents failed to remember the number of days they missed work for sickness. The mean work-day loss was around one and half months a year ($\sigma = 43.96$ days) for the BRHKs, primarily due to travelling to mainland China. They incurred additional work-day losses for travelling home. Although travelling home incurs costs, it also provides two advantages: migrants got their visas extended upon return; and second, migrants were able to reconnect with their families.

Apart from these, 17 of the BRHKs were charged on various offences and were detained, and spent on average 7.12 days in custody. They remained unemployed around 80 days a year. The highest percentage of the migrants remained unemployed between 80 and 90 days. The BRMs experienced relatively fewer work-day losses than the BRHKs for visiting their home country, although BRMs spent significantly higher number of days in jail. Ultimately, the number of days that the BRHKs spend not working were not significantly different than those of the BRMs.

'My home, my sweet home'

A number of studies have considered the migrants' living quarters as a significant element of their decision making with respect to migration (Abdul-Aziz 2001). This study considers the current residential condition (living arrangement i.e. whether they live in single room or shared or in a mess) and the rental arrangements of the respondents. Data show that in the case of the BRHKs, around one-third (32 percent) lived in shared flats and the rest in shared rooms. In Hong Kong, in most cases, renting a house required the help of an agent and usually a Hong Kong Identity (HKID) card. Around 34 percent of the BRHKs were able to rent rooms by using a proxy, often their local girlfriends, who have an HKID. Meanwhile, 33 percent found their accommodation through Bangladeshi friends, or by some shared living arrangement.

Generally, when a labour contract is negotiated, even for unskilled workers, the package includes free accommodation. As such, most of the BRMs were promised free accommodation during their tenure in-country. For many of the migrants, it seems that the provision of housing was a key factor for them when considering their move. This is because having the cost of accommodation covered would increase their estimated savings.

Most of the BRMs without a contract also managed to have free accommodation. This study has shown that some 20 percent of the BRMs were living in very poor conditions. They were either packed in unventilated container vans at their workplace or lived in a small corner inside the factories in which they were employed. While some lived in shared flats, these arrangements had to be made at their own expense, with the help of friends, and occasionally through their employers. However, most of the BRMs intended to arrange accommodation by themselves. They said the cost of accommodation in Malaysia is cheap and, therefore, they could have arranged for better living conditions, had they been given a choice.

This section explores who extended help to rent the accommodation of the migrants in host countries. The importance of this aspect lies in the fact that the level of the assistance received essentially indicates the legal position of the migrants. The renting processes in Hong Kong and Malaysia are distinguished by the requirement of a guarantor to sponsor any non-HKID holder in renting a house or apartment in Hong Kong, a condition which does not exist in Malaysia. Apart from acting as guarantor, the migrants require assistance to find a suitable house to rent and negotiate rates. In this case, the many of the BRMs received assistance and none of the BRHKs received such help from their employers. Most BRHKs reported having received assistance from their friends and agents.

Table 6.3 Conditions of accommodation

	Hong Kong		Malaysia	
	n=56	%	n=70	%
No reply	–	–	2	2.8
Living 'rough'/in forest*	–	–	2	2.8
Shared flat	18	32.14	23	32.9
Shared room or mess	38	67.86	4	5.71
In a container	–	–	25	35.7
Inside a factory	–	–	14	20.0
Total	**56**	**100.0**	**70**	**100.0**

Note: '–' indicates not applicable; * They were living in the forest because their house was regularly raided by the police.
Source: Author's field data, 2004–2006.

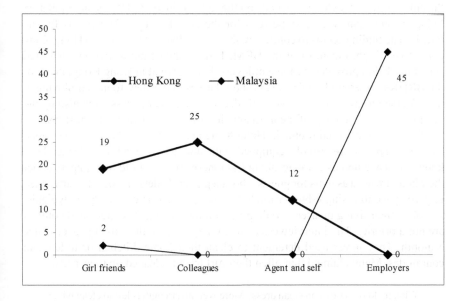

Figure 6.2 Persons extended help renting a house
Source: Author's field data, 2004–2006.

This section has analyses of the comfort and discomfort factors in the living place which included how congested and hygienic are the living. The study shows that a few of the BRHKs live in a single room and the rest in shared rooms. However, a few others said that some of their friends rent a single room, not so much for comfort, but to have a place to meet their girlfriends. A few others explained that renting a single room was something they could do, because they do not send money to their families anyway. Therefore, it was not always true that the ability to rent depends on the level of the income.

For the BRMs, many arranged their accommodation by themselves. They live in more spacious place as house rent is relatively cheaper. But the living conditions of the accommodations provided by employers are poor in every respect. Mohon (a shoe factory worker at Selangor in Malaysia) reported:

> We are labourers – coolies. We don't require a place to sleep. It is for the boss ...
> Who cares? ... We 11 people sleep in a tiny basket ... Our employer has never
> peeped into that [to see] if we are still alive.

A significant percentage of the BRMs (39 percent) shared room with at least five room mates; around one-third with 6–10. Many others (30 percent) shared room with 11–15 boarders. A significantly higher percentage of the BRMs lived in a single room as compared to the BRHKs (P=0.000). It describes the situation where the amount of rental charge depends on the level of income. The mean population living in one room was 2.55 persons for the BRHKs and 7.21 persons for the BRMs. The finding seems to contrast with the fact that rental charge in Hong Kong is relatively high. For most of the BRMs had no choice but to live in congested accommodation provided by the employers. Upon arrival in Hong Kong, many of the BRHKs stayed with their friends and relatives. After they found employment, they decided to live on their own. This decision was later necessitated also by the need for privacy as many of them became involved in romantic relationships.

Typically, the rented rooms in Hong Kong are not bigger than 20 square feet, and these spaces are usually equipped with double- or quadruple-decker beds known as 'cage homes', 'loong-uk' in Cantonese. Heaps of old newspapers under their beds are used as mats for guests who sleep over. Toilets are loosely partitioned to give space to a single-burner kitchen, a setup which is obviously unhygienic. A look at their living quarters would give an impression that hygiene and comfort are not a priority. The interviewees claimed they cleaned their living spaces once a month. They never used detergent to clean their home, kitchen or toilets and commodes. Many claimed that often their girl friends cleaned up their beds.

> What to do with tidy and clean dress? Moreover, no strength is left to clean when
> we are back from work. (Habib, a migrant worker, Hong Kong)

The above two excerpts describe similar phenomenon but from very different standpoints. The former does not reflect any grievances while the latter speaks out of frustration. Abdul-Aziz (2001: 17) had similar findings about the accommodation of the BRMs, explaining that the workers were housed in nothing more than dilapidated shacks. Conditions inside can be quite deplorable; they are cramped, dimly lit and poorly ventilated. For most of the BRMs, day and night shift workers took turns sleeping in the same bed.

Table 6.4 Number of boarders living in single room

Number of boarder	Hong Kong		Number of boarder	Malaysia	
	n=56	%		n=70	%
1	3	5.36	<5	27	38.57
2	25	44.64	6–10	22	31.43
3	22	39.29	11–15	21	30.0
4	6	10.71	–		
Mean	2.55			7.21	
SD	0.761			4.093	
Significance*			P=0.000		
Total	56	100.0		70	100.0

Note: '–' indicates not applicable; * BRHKs vs BRMs.
Source: Author's field data, 2004–2006.

Travelling to and from the workplace consumes time and money. 'Distance' in this study refers to the time required to get to work from the workers' places of residence because there was no single workplace for most of the respondents. The study did not consider the physical distance for two reasons: First, the BRHKs do not have any defined place of work – workers were most often driven to work places in the employer's cars. Second, many of the BRMs lived within factory premises. A few of the BRMs built temporary dwellings in nearby forests, and this case was observed in Johor Bahru of Malaysia. Data show that the highest percentage of the BRHKs lived farthest from their work. For the BRMs, around half stayed within the shortest range of distance from their work. They either lived near the factories and or within the factory premises. This study includes analysis of the place in which the respondents they wait in order to get job offers.

   The aim of this analysis is to understand the relationship between accommodation costs and the income level of both the BRHKs and the BRMs. The significance of this analysis lies in the fact that the relationship between migration and income often reflects the living conditions of the migrants (Krieger 2004). Accommodation cost is considered a part of the gross income earning, which is an important determinant in migrants' choices in terms of type and location of accommodation. This study has confirmed that the difference in the accommodation cost between the BRHKs and the BRMs was too great to make a meaningful analytical comparison. Therefore two tables present the situations in two countries separately under the same heading. The study shows that rental charges are significantly related to the level of income (P=0.000) for both the BRHKs and the BRMs. The rental charge increases with the growth of income level.

Table 6.5 Accommodation charges in relation to the level of income

Monthly income HK dollars	Monthly rental				Total
	5000–7000	7001–9000	9001–11000	11001–13000	
400–600	5 (8.93)	–	1 (1.79)	–	6 (10.71)
600–800	14 (25.0)	6 (10.71)	5 (8.93)	–	25 (44.64)
800–1000	7 (12.5)	–	7 (12.5)	–	14 (25.0)
1000–1200	2 (3.57)	–	3 (5.36)	3 (5.36)	8 (14.29)
1200–1400	–	–	–	1 (1.79)	1 (1.79)
1400–1600	–	–	–	2 (3.57)	2 (3.57)
Significance			P=0.000		
Total	**28 (50.0)**	**6 (10.71)**	**16 (28.57)**	**6 (10.71)**	**56 (100.0)**

Monthly income Malaysian Ringgit	Monthly rental						Total
	0–50	50–100	100–150	150–200	200–250	250–300	
0–500	2 (2.86)	2 (2.86)	–	–	–	–	4 (5.71)
501–1000	11 (15.71)	2 (2.86)	1 (1.43)	1 (1.43)	–	3 (4.29)	18 (25.71)
1001–1500	25 (35.71)	3 (4.29)	11 (15.71)	–	–	2 (2.86)	41 (58.71)
1501–2000	4 (5.71)	–	–	–	2 (2.86)	–	6 (8.57)
2501–3000	1 (1.43)	–	–	–	–	–	1 (1.43)
Total	**43 (61.43)**	**7 (10.0)**	**12 (17.14)**	**1 (1.43)**	**2 (2.86)**	**5 (7.06)**	**70 (100.0)**

Note: '–' indicates not applicable; Figures in parentheses indicate percentages.
Source: Calculated from author's field data, 2004–2006.

Adaptation strategies in rationalizing post-migration decision

The preceding chapters have provided both qualitative and quantitative analysis about rationalization of migration decision to support the research argument that expectations of economic gain and a better future prompted these workers to make their decisions to migrate and work overseas but that these hopes remained, in most cases, unrealized. In this section, I pursue several questions in order to understand how Bangladeshi migrant workers rationalize their migration decision. Based on interview data, I identify the forces inducing migration decisions and specify why it is that the migrants often adhere to their decisions for many years.

In the contemporary research literature on this topic, adaptations and assimilation are seen as significant variables in the 'rationalization' process (Jackson 1986). Therefore, in this section, I analyze different adaptation strategies such as language, companionship, patterns of leisure time usage, and a miscellany of other challenges migrants commonly encounter abroad. In addition, migrants perceive that the level of access to welfare systems of the host countries such as healthcare service is a significant consideration for continuing to stay in the host country.

Adaptation into the host country is a significant issue in rationalizing migration decisions for the migrant workers. From a psychological point of view, rationalization is the process whereby a mind becomes defensively blinded to certain understandings (see Elster 2000) and according to Krieger (2004), the initial phase of the migration process is characterized by disinterest in migration followed by developing an inclination to migration and integrating the possibility of migration into a range of potential alternative actions.

Migration for manual work is, in itself, a source of vulnerabilities (Waddington 2003) which increase when the migrants stay in the host countries as 'illegal' workers – they include threats of imprisonment, sexual harassment, rape, physical and psychological mistreatment, verbal abuse and unplanned pregnancies (Ullah 2005). Illegal migrant workers either consciously assimilate into the host country's culture, or they must adopt a more clandestine lifestyle, somehow disguising themselves and their activities from the authorities.[9] In most cases, the migrants are seen primarily as outsiders, and they hold onto this status until they can adapt and assimilate into the host society. Theoretically, the fundamental issue of inclusion and exclusion is based on differentiation between human groups. In the new economic order, subsistence families break down, and are replaced by participation in national and international markets. Individuals who possess the characteristics necessary to fit into global markets, whether for labour, capital or cultural goods, are included into the global order as citizens, with civil, political and social rights. Individuals who are excluded may be denied even the most basic rights, such as the right to work and food security (Castles 1998). Therefore, in these circumstances they need to explore means and strategies for adaptation and assimilation into the receiving countries (Landau 2004). Migrants adopt different kinds of strategies to this end which might vary from individual to individual and country to country. However, the level of assimilation depends on the socio-

---

9 During the crackdown (which has been taking place in Malaysia with the goal of eliminating illegal migrants) many fled to the jungles to escape arrest and repatriation.

cultural and political conditions of the host countries.[10] Therefore, there are differences in the range of strategies adopted by the BRHKs and the BRMs to adapt and assimilate to their respective host countries.

This section discusses how local language is used as a part of migrants' adaptation strategy into the local society. It is often the case that the migrants' native tongues differ from the languages spoken in the host country. Deficiencies in the ability to communicate with locals are likely to be a major factor that constrains the earnings of migrant workers (Ullah 2010). The ability to communicate with the local population is an important factor for a migrant because language proficiency is often used by employers as a screening device for recruitment (Dustmann 1994). As a part of cultural assimilation, many of the respondents agreed that learning the local language was a common strategy for adapting to the receiving society. However, assessing the language skills of the respondents, predominantly based on self assessment, is problematic for information used for statistical purposes. The reason is that respondents have different perceptions about thresholds in language capacity (Dustmann 1994).

Most of the respondents in this study had learnt local languages. The BRHKs had learnt Cantonese to communicate with the local employers and with the police as and when required. In order to surmount the language barrier, they first learnt Cantonese from their compatriots, relatives, friends or language teaching schools. Data show that a vast majority of the respondents reported speaking some Cantonese and one-third claimed to speak it fluently. The BRHKs stated that both Cantonese and Mandarin were relatively difficult to learn. Most BRMs spoke Bahasa Malaysian. Around half of them also spoke Hindi, although, the BRMs stated that knowing Hindi was of little assistance. One-third spoke Bahasa Indonesian. As regards to their proficiency in Bahasa Malaysian, an overwhelming majority (84 percent) claimed to be fluent. The level of proficiency does not always depend on their need to find a job easily: it also depends on their willingness to learn and the length of residence in the respective country. The majority of respondents (BRHKs and BRMs) have little command of English.

The respondents expressed the belief that knowledge of the local language provides access to a broader spectrum of jobs, as well as a larger set of possibilities to occupy better paid positions. Apart from this, language skills tend to limit vulnerability to exploitation (Dustmann 1994). While many factors contribute to psychological stress, language plays an important role in mitigating stress by removing barriers to communication that may compound feelings of isolation. Communicability influences healthcare-seeking behaviour such as: under-reporting, poor explanation of health problems and symptoms, inappropriate

---

10   DeVoretz also questions whether there exists a set of general economic principles that can guide migrants in this assessment? Or is the world so idiosyncratic that each state has its own implied social welfare function such that economics cannot guide us? Social and economic integration go hand in hand. He also asks what measures are available for the migrants to self-assess their degree of integration (DeVoretz 2004, Bauböck 2005).

diagnoses and the capacity of immigrants to comply with treatment regimens (Carballo and Nerukar 2001). This section has further examined how language is used in the adaptation strategies of migrants.

Table 6.6 Language proficiency (multiple response)

BRHKs (n=56)	Fluent	Very good	Good	Below standard	Total
Bengali	56	–	–	–	56
English	–	19 (33.93)	–	31 (55.36)	50
Cantonese	–	21 (37.5)	13 (23.21)	9 (16.07)	48
Others (Malay, Korean)	–	–	3 (5.36)	5 (8.93)	8

BRMs (n=70)	Fluent	Very good	Good	Below standard	Total
Bengali	70	–	–	–	70
English	3 (4.29)	1 (1.43)	5 (7.14)	61 (87.4)	70
Malay	59 (84.29)	7 (10.0)	3 (4.29)	1 (1.43)	70
Bahasa Indonesian	1 (1.43)	6 (8.51)	3 (4.29)	11 (15.71)	21
Hindi	9 (12.86)	15 (21.43)	2 (2.86)	3 (4.29)	29

Note: '–' indicates not applicable; Figures in parentheses indicate percentages.
Source: Author's field data, 2004–2006.

The collected data show that an overwhelming majority of both the BRHKs and the BRMs thought that capacity in the local language helped them to avoid police arrest or to obtain a release if they were arrested. Around 93 percent of BRHKs and 91 percent of BRMs found that local language capacity helped them to obtain work. Interestingly, migrants were sometimes careful in using language skills. For example, people who were proficient in the local language sometimes pretended not to be, with BRHKs feigning ignorance in order to escape or avert interrogation by the police; while the BRMs said Malaysian police sometimes used language proficiency to identify long-term migrants and target them for repatriation.

This section shows the significance of companionship in the adaptation strategy of the migrant workers. Companionship can often lead to romantic liaisons. Migrant workers consider having partners or companions as an important adaptation strategy. A couple living together without the bond of marriage is traditionally frowned upon in Bangladeshi society, for both social and religious reasons. However, while abroad, Bangladeshis often become involved in such relationships.

These companions and friends inevitably become a critical factor in migrants' stay because they extend support to the migrant workers in a number of ways. For example, illegal migrants do not possess the Identity Card essential for renting a house and getting a telephone connection, especially in Hong Kong.

Their companions can substitute their own cards on behalf of their boyfriends. Some migrants also opened bank accounts with the help of their girlfriends. The study shows that almost all the BRHKs reported having girlfriends. Interestingly, about two-thirds of them had multiple girlfriends, most of whom were from the Philippines and Indonesia, working in Hong Kong as domestic helpers. Kamrul (age 39), a migrant worker in Hong Kong stated that it was easy to have girlfriends in Hong Kong. He added that one can even have ten girlfriends if he wants. Many of them said that 'thousands' of Filipino, Indonesian and Thai women work in Hong Kong and thus, finding girlfriends from among them was not difficult. The added factor of isolation and distance from home make these migrants – both men and women – long for companionship.

The data further show that the BRHKs with a partners/girlfriends were mostly unmarried: The number of unmarried migrants with girlfriends was significantly higher than those who were married (P<0.001). Married migrants who did not have girlfriends reported that they considered having girlfriends as an injustice to their wives back home. However, there were exceptions. For example, Jashim, a married migrant with a girlfriend said, 'we are human beings, still young. My wife is at home and we did not leave our carnal desire at home'. This indicates that respondents' physical separation from home creates a desire for companionship. Interestingly, there is a mutual understanding among the migrants that they would not make the issue of having girlfriends known at home.

An overwhelming majority of the BRMs (69 percent) had girlfriends: Approximately one-third of the married (at home) BRMs had girlfriends, and among the unmarried men, around one-third had girlfriends. Therefore, unlike the BRHKs, there was no significant difference between the married and unmarried men in their propensity to have girlfriends. There were also different views on having girlfriends between the BRHKs and the BRMs. To the BRHKs, the availability of single women in Hong Kong prompted them to have girlfriends, while the BRMs are inclined to having extramarital affairs because they gave up hopes to be bale to return home soon.

Finding ways to spend spare time is one of the major concerns for most of the study respondents in rationalizing their decision. However, the pattern of spending their spare time among the BRHKs and the BRMs varied widely because of the varied nature of their work and legal status.

Data show that slightly more than one-third (34 percent) of the BRHKs spent their leisure time with their girlfriends. Many others (around 40 percent) opted not to reply. Many of them, I felt, seemed reticent to reply and some others were unsure of what to say and as a result neglected to respond to the interview questions. Although during holidays, Hong Kong turns into a city of migrants because of their lively presence everywhere (Law 2002) and Kuala Lumpur is quite similar, the majority (61 percent) of the BRMs report spending their spare time at home watching television and movies on a VCR so as to avoid the risk of being caught by the police, which might entail paying a bribe to obtain a release. Some other

BRMs went out for Tablig[11] activities, while some sold phone cards in their spare time. When I visited the houses of the Bangladeshi workers in Malaysia, TVs and VCRs were continuously playing. It was safer for them to stay at home. Many migrants had no errands to perform during their spare time.

Table 6.7 Respondents with partners by marital status (cross table)

Hong Kong		Having partners			Total
		Yes	No	NR*	
Marital status	Married	9 (16.07)	10 (17.86)	–	24 (42.86)
	Unmarried	31 (55.36)	3 (5.36)	3 (5.36)	29 (51.19)
	NR*	3 (5.36)	–	–	3 (5.36)
χ^2 significance		P<0.001			
Total		*43 (76.79)*	*13 (23.21)*	*3 (5.36)*	*56 (100.0)*
Malaysia					
Marital status	Married	20 (28.57)	8 (11.43)	10 (14.29)	38 (54.29)
	Unmarried	28 (40.0)	1 (1.43)		32 (45.71)
χ^2 significance		P<0.006			
Total		*48 (68.57)*	*9 (12.86)*	*13 (18.57)*	*70 (100.0)*

Note: '–' indicates not applicable; * No response; Figures in parentheses indicate percentages.
Source: Author's field data, 2004–2006.

In most cases it is the employers who grant leave and holidays. However, migrant workers who work on hourly basis generally decline holiday leave because they prefer to work for long hours to earn more. In Hong Kong, many migrants wait for employment offers on weekends and holidays. Enjoying holidays applied to only 12 of the BRHKs (21 percent) who had a contract with employers. Data show that one-third (33 percent) of them reported enjoying holidays on Sundays, and half of them enjoyed according to the wish of the employers and some others (17 percent) at the end of the year. A majority of the BRMs enjoyed a holiday on Sundays, another one-third enjoyed holidays when employers wished and a few others (12 percent) enjoyed end of year holidays. As in Hong Kong, a few intended to work on their holidays in order to earn overtime. None of the respondents had control

---

11   Tablig Jamaat concentrates on preaching and teaching the rituals of Islam. The Tablig followers execute the programme by moving about far and wide seeking and imparting knowledge and spreading the lofty injunctions of Islam, adhering to them strictly.

over the decisions regarding their holidays. (Question: Do you get days off? Only positive answers were considered; Question: When do you get days off?)

The following sections investigate the various challenges in adapting to this aspect of the host country.

In adopting different adaptation strategies, migrants encounter a number of challenges or frustrations. Frustration is a human emotion that occurs in situations where one's goals remain unrealized. It is also regarded as a useful indicator of problems in a person's life (Elser 2003: 7). The migrants take on a number of adaptation strategies to cope with the host environment but, some circumstances like frustration and dissatisfaction cannot always be controlled by adopting these strategies.

The study shows that the expressed levels and sources of frustration varied widely between the BRHKs and the BRMs. The BRHKs were frustrated with new 'competitors' coming into the labour market (Münz 2004, *Daily Ittefaq* 2007). Long-standing migrants tend to compare their stay in Hong Kong under British and Chinese rules. According to them, the former was better. They explained that, under British colonial administration, demand for labour was high, consequently the wage rate was higher, competition for jobs was less intense, stay permits were lengthier and easier to obtain, and police surveillance was less stringent.[12] Bangladeshi migrant workers observe that before the handover to the Chinese in 1997, the job market was better because there were not so many Chinese from the mainland crossing the border seeking work. They think, and rightly so, that this may be attributed to new immigration policies after the handover in 1997 (Lee 2006). These new migrants have an advantage as many of them can speak Cantonese fluently. Of late, migrant workers from Nepal, India and Pakistan, have also become more active competitors for the same categories of job sought by the Bangladeshis. Babul (aged 37), a Bangladeshi worker in Hong Kong since 1994, laments: 'now we need to keep waiting for a whole day to get one work offer, while we would get more than one [work offer] before the handover'. The following statement provides a sense of how Bangladeshi migrant workers think about their competitors.

> ... when a Hong Konger approaches them for a work offer, the mainlanders rush to them and the employers turn us down. Thus, they snatch away our work offer. This competition has lowered the wage rate as well. (Awal)

This section is not meant to simply compare the situation before and after handover, rather it depicts how the migrants adapted themselves to the labour market after the condition turned highly competitive. It is plausible that migrants are always a disadvantaged group in the labour market and the lower rates of

---

12   As mentioned earlier, a visit visa was for three months, before 1997 after which it was for two weeks, and then, after 11 December 2006, none.

participation in jobs by migrants could be attributed to factors such as the similar ethnic background between the Mainlanders and the Hong Kong employers.

In Malaysia, Bangladeshi migrants were also frustrated for a number of reasons. The BRMs state that the Indonesians have occupied the Malaysian labour market, and become competitors for the Bangladeshi migrant workers in Malaysia. The BRMs said that after the bilateral labour export agreement between Bangladesh and Malaysia became ineffective in 1999, Bangladeshi migrant workers were viewed as undesirable elements in Malaysia. They subsequently pointed out that police harassment has increased and exploitation by employers worsened.

The legal migrants and illegal workers: New terminologies developed

Amirul, a Hong Kong resident, points to a problem faced by many of the interviewees when he observes, 'I am a legal migrant, but not a legal worker'. As an adaptation strategy, the BRHKs consider being a legal migrant important because they are aware that as long as they legally stay in Hong Kong they can look for employment, even if being employed is against the provisions of their visa. Therefore, legal residency is their primary concern because staying illegally in Hong Kong is more difficult than in Malaysia. For this reason, they apply particular strategies for keeping their stay permit in order.

As noted earlier, extending a stay permit while in Hong Kong is not possible. They need to exit in order to seek an extension of their stay when they return. As mentioned, they prefer to go to Shenzhen, Macau being their second choice, as the cost is greater. On return to Hong Kong, most of them are permitted to stay seven days on an average. But not all are so lucky, getting less than seven days, while some are denied and must return home, perhaps returning with another passport issued in a different name. The migrants stated that they could bribe the police. Besides, they use fake visa stickers that are not easily detectable without the help of a machine.

The migrants call a 'denial of extension' as 'report khaisee' – i.e., caught and deported. To any migrant, this is a source of embarrassment and difficulty should they return to Hong Kong. The record of denial in the same passport would call for more than the usual routine questioning at the immigration counters for their return. They call the 'passport' an 'elephant'. To keep the passport in order they require to exit which involves a lot of money. They gave this instance since nurturing an elephant requires a lot of money. Shob taka ai hati palte shesh hoi – i.e., 'The elephant sucks all the money' – which is a typical reply of the BRHKs when asked how they stay in Hong Kong. As mentioned before, 93 percent of the respondents held tourist visa, while the rest claimed not to have any visa or any passport. Data further show that 34 percent of them came to Hong Kong more than once and 66 percent between 6 and 10 times.

The migrant workers have a self-developed terminology which grows as they seek to enhance their adaptation strategies. During the ethnographic survey, I found migrants using a number of words and phrases that were unfamiliar to me. When asked why they used these terms, they replied 'Eda amgo bhasha' – i.e. 'This is our own language!' One example is the term they use for the police, 'Mamu'. This is a Bengali word that in English means 'uncle'. This word is very useful during police crackdowns. Migrants alert the other friends by using this word so that the police do not understand their message.

Policemen carry out sudden raids in places known to be popular spots for illegal workers. Sometimes such raids result in a high number of illegal migrants arrested. However, the migrants are often spared, according to many of them. Sometimes the police ask why the men were waiting on the street. They usually respond that they are waiting for friends to meet them. If the migrants cannot produce valid travel documents during police spot checks, the migrants are arrested, although their compatriots may help them by fetching their passports, thereby obtaining their release from custody. In recent times, the police have begun to collect and preserve their fingerprints or other biometric identifiers for future reference (on biometrics, see Thomas 2005). When they are arrested for any offence, the Hong Kong police check to see if they had been apprehended before. If so, they are sent to jail for six months in the first instance, while repeat offenders get an 18 month jail sentence. In a letter dated 5 May 2005 from the Correctional Services Department (CSD) in Hong Kong, information was provided regarding the offences normally committed by Bangladeshi migrants (ref: 87 in CSD31-50-pt2). Most are immigration-related, including breach of condition of stay (40 percent), the making of false statements to immigration officers (35 percent), and the use of forged travel documents (25 percent). They receive sentences from seven days to 11 years of imprisonment.

There is an interesting similarity among the BRHKs and the BRMs in that both groups call the police 'Mamu'. ('Soto mamu' – i.e., younger uncle – refers to the patrol police, and 'boro mamu' – i.e., elder uncle – refers to the immigration police.) However, the Hong Kong police cannot be bribed, as opposed to the Malaysian police, who ask for bribes. There are instances where bribes can be arranged on credit, and migrants with no money promised to pay the officer after they receive their salary. However, the immigration police (the 'boro mamus') in Malaysia normally do not seek bribes.

There are some other means by which the migrants adapt, although the strategies adopted by the BRMs are different from those of the BRHKs. Due to Kuala Lumpur's strategic plan to become a modern industrial nation by 2020, and a change in demand from unskilled to skilled workers, Malaysia, according to analysts, is getting tough on illegal workers (*Migration News* 1997). Malaysia will soon require migrant workers from several countries to pass a course on Malaysian law and culture before they can be issued work permits.

A 'stay permit'[13] for Bangladeshi migrants does not necessarily permit the migrant to work. In Hong Kong, violations of the rule of employing illegal workers or accepting employment without a work permit carry huge penalties. The following presents information on some strategies used by the BRHKs and the BRMs to continue working. The four major categories of strategies adopted to stay are hiding, bribing police, seeking help from employers and confinement.[14] While one-third of the BRMs reported bribing police, none of the BRHKs reported bribing Hong Kong police. Abdul-Aziz (2001) agrees that tempted by the opportunity to make quick money, some dishonest police officers often take advantage of their position to extort money from these people in return for acquiescence and for this reason police raids tend to occur on paydays. Around one-third of the respondents were stopped by the police on an average of four times, ostensibly for spot inspections of documentation, but actually for money (Abdul-Aziz 2001).

A number of respondents reported using fake work visas, which cost between RM500 and 3000. Respondents explained that some migrant workers make fake visa stickers by scanning an original. The scans are of such quality that patrol police cannot identify them as fake without a detector machine.

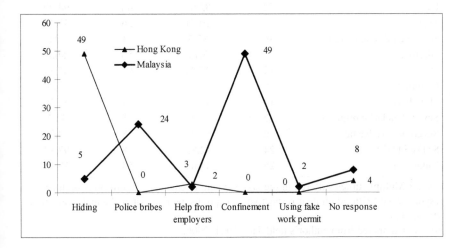

Figure 6.3 Strategies adopted continue working (multiple responses)
Source: Author's field data, 2004–2006.

---

13   Important to mention that Bangladeshi passport holders were granted a 3-month stay permit upon arrival the airport before 1997, after which it has been reduced to 14 days. But from 11 December 2006, no permit on arrival is allowed.

14   Hiding and confinement hold two distinct implications. Confinement refers to a situation where employers confine their workers to avoid police gaze. Hiding refers to hiding their purpose of staying in Hong Kong and situations where the illegal migrants try to hide themselves i.e. by staying in a place where police surveillance is relatively low, such as within the factories by locking them from the outside, or in a remote house.

As mentioned before, access to a welfare system is an important issue that the migrants consider in rationalizing their decision to migrate. There has been a great deal of debate about the rights of non-citizens which revolve around the problem of social exclusion. Both legal and non-legal migrant workers have limited access to social services in many countries; and this is the case for Bangladeshi migrant workers in both Hong Kong and Malaysia. There is an ongoing debate on the rights of access to welfare services entails health, disability, and retirement and old age benefits (Maharaj 2004). This section intends to examine the different kinds of illnesses they suffer including occupational illness and the pattern of seeking treatment, with data representing [respondents' perceptions of] the situation one year before the interview date.

Table 6.8 Diseases suffered in a year* (multiple response)

Morbidity	Hong Kong		Malaysia	
	n=56	%	n=70	%
Fever	52	92.83	66	94.29
Typhoid	6	10.71	26	37.14
Severe colds and coughs	49	87.5	21	30.0
Diarrhea	4	7.14	56	86.0
Dysentery	13	23.24	61	87.14
Rheumatism	10	17.86	5	7.14
Skin disease	17	30.36	31	44.29
Severe headache/migraine	42	75.0	25	35.7
Severe stomachache	36	64.29	58	82.86
STDs/STIs**	24	42.86	68	97.14
Unknown	25	44.64	54	77.14

Note: * Multiple times suffered in the last one year (frequency of sufferings/frequency of respondents); **Based on their respective perceptions, pain in urinal channel, scars on the sex organ, etc.
Source: Computed from author's field data, 2004–2006.

Health Many respondents reported carrying some of the illnesses listed in the Table below, including rheumatism and skin diseases, when they migrated. Data show that majority of the migrants suffer from various types of illnesses while frequency of sufferings and pattern of morbidities vary between the BRHKs and the BRMs. Among the BRHKs and the BRMs, the highest percentage report suffering from fever (93 and 94 percent, respectively). An overwhelming majority (86 percent) of the BRHKs suffer severe cold and cough, while only one-third of the BRMs suffer from the same complaint. This is because of the nature of their work, for example, the BRMs spent longer hours in the factories than the BRHKs.

In addition, the BRMs were often forced confined in the factories. Also the reported unhealthy working conditions in the factory contribute to the increasing health complaints.

Occupational illness   Occupational hazard or illness is a term for a common form of injury which ranges from minor irritations to major accidents affecting migrant workers. Migrants' working conditions are such that they are always more prone to such hazards than the resident workers, especially those who work in construction, mining, heavy manufacturing industry, and the agricultural sector, where safety measures are insufficient. The vulnerability to unsafe and dangerous working conditions includes exposure to occupational injuries, chemical, and other toxic materials or the effects of sustained work in substandard surroundings (Carballo and Nerukar 2001). Lax enforcement of labour standards for the employers of illegal migrants have contributed to aggravating their occupational vulnerabilities (Prothero 2001).

Further, this study highlighted the categories of work for which the migrants suffered inordinate health risks. BRMs suffered more occupational hazards than the BRHKs, indicating that working in Malaysia was more dangerous for migrants than working in Hong Kong. Lack of awareness in the proper use of chemicals, poor access to sanitation facilities, and ignorance of the need to maintain good standards of personal hygiene contributed to the high frequency of skin disease among the BRMs. Though the number of occupational injuries remained low during the periods surveyed, the injuries could be due to other causes such as cuts and burns suffered to extensive areas of hands and other parts of the body. There was high incidence of head injuries reported by employees. These ranged from minor external injuries such as bruises, cuts and burns to fractures to more serious injuries.

Most back injuries are caused by the heavy load of the basket saddled to the back and carried by both the BRHKs and the BRMs. Injuries to the spine may result in severe back disorders with long-term effects. Occupational injuries to feet and toes were also high among the respondents, primarily cuts and lacerations due to the improper use of tools and equipment, slips and falls, and the failure to use safe footwear.

Healthcare-seeking pattern   Part of the social exclusion suffered by the migrants means that they do not have access to the welfare system in their host countries. This becomes especially clear when the healthcare-seeking behaviour of the interviewees is taken into account. The BRHKS and the BRMs are generally averse to seeing a doctor for afflictions related to their work. The reluctance to visit a doctor is more evident among the BRMs than the BRHKs, and access to public hospitals for both the BRHKs and the BRMs is limited.

Figure 6.4 shows two noteworthy aspects: First, there is the difference in the frequency of visits between the BRHKs and the BRMs; and second, there is the difference in frequency of visits to private and public hospitals among the respondents from both groups. A higher percentage of the BRHKs (57 percent)

visited a hospital as compared to the BRMs (41 percent). The BRMs stated that
they were reluctant to visit a hospital because of the lack of money, their lack of
legal status and what they identified as rude behaviour by the hospital staff. In
explaining this, most of the BRMs reported that doctors and other staff treated
migrant workers badly. The migrants were not given any emergency or priority
service when required. A number of the BRMs complained that they visited the
hospital with serious wounds the hospital staff would seem to be busy with other
patients, and they had to wait long time in the queue. The BRHKs were averse to
visiting a hospital primarily because of the expense of medical treatment. A higher
percentage of the BRHKs visited private clinics as compared to the BRMs. None of
the BRHKs reported 'misbehaviour' by Hong Kong doctors and hospital staff.

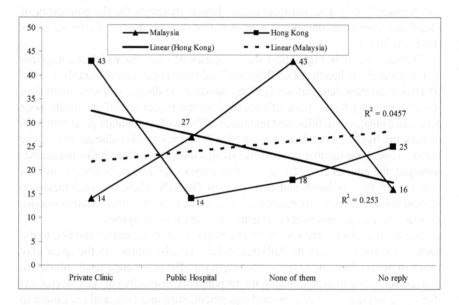

Figure 6.4 Healthcare seeking behaviour
Source: Author's field data, 2004–2006.

Mitigating psychological stress

Networking does not end once the migrants arrive at their destinations; rather
it continues in other forms. The issue of post-migration networks has not much
to do with migration decisions themselves, however it is a significant factor in
the migration cycle and this influences migrants' decisions to prolong their stay
abroad. This section discusses how the migrants maintain communications with
the families they left behind. Through communication, migrant workers remain
informed about the overall state of the family, especially how their remittances

are being used at home. This aspect needs to be taken into consideration as post-migration communication contributes to the emotional state of both the migrants and their families. Most of the respondents maintained regular contact with their families in Bangladesh by multiple modes of communications. However, there was a difference between the BRHKs and the BRMs in the mode of contact used. A few of them (BRHKs) claimed that they were in touch with their families through email whereas none of the BRMs claimed to have used email. Rather, most of them made phone calls[15] to their relatives, although, they felt that it was expensive. They also sent letters through the postal service. They preferred to write letters as, according the respondents, it is cheaper and easier.

> … [I have] no concern about my husband because he phones us regularly. He needed a document last month; he called and instructed, we sent the document to him. It was so easy. (Shafina, wife [in Barisal, Bangladesh] of a migrant in Hong Kong)

Like Shafina, many others underscored the importance of communication in migration. Most of the BRHKs talked to their family members over the phone, while a few used email, as it was the cheapest mode of communication. They used email from cyber cafes or at Mass Transit Railway (MTR) stations for free. Both BRHKs and the BRMs were aware of the economic conditions of their families, they know about their children and how the money was being spent. Many respondents said that a visit to Bangladesh would be much better than sending email or letters in order to oversee how their money is spent. However, it is expensive to visit. The BRHKs do not visit as frequently as they would like to, and most of the BRMs who want to visit do not do so for lack of money and for fear of being denied re-entry. The difference in the frequency and modes of communication between the BRHKs and the BRMs lies in the extent of the facilities available in the respective destinations, such as in cyber cafes and MTR stations.

Data show that the BRHKs visited their home country at least 6 times during the total period of their stay in Hong Kong (mean residence is 4.84 years). Most of the BRHKs did not intend to go back home until they achieved their 'goal', which was first to earn back the amount of money they spent on financing their migration, and second to save for the future. Their monthly income could finance only one return trip from Hong Kong, but the less frequently they visit, the higher their savings are. A question may be asked: why did they visit home so many times? Some of the replies expressed remorse, some with sighs and some with wrath against the Bangladesh government for not taking care of their concerns.

---

15   Mobile phone networks have percolated beyond poverty boundary in Bangladesh, even to the doors of the poor in the remote rural areas. It is a recent and surprising advancement in Bangladesh. In the last few years (less than five), the network reached almost all the districts. The migrants admitted that they are now the beneficiaries of this advancement. Therefore, easy communication has facilitated migration.

The BRHKs said that they only go back home under some forced conditions, such as being repatriated by the immigration department, or having their passports 'report khaisee' (i.e., reported), the migrants' euphemism for a 'denial of extension' – which caused them to be caught and deported. Some returned because of the demise of close ones, and so forth. Migrants did not report their periodic returns as pleasure trips. Some of them said that they had to go back home when they were denied extension of visa on return to Hong Kong. Similarly, they sometimes visited home to get their passports renewed or reissued.

My data show that the majority of the BRMs (66 percent) never went back to the country after their first arrival in Malaysia. Only about 16 percent of them visited their families once and 19 percent only twice during their total period of residence (average residence period is 8.93 years). The most articulated reasons for not visiting homes were that their travel documents were not in order; a few had no money to go back, and others said that going back home would be of no use as they had no errands to do there.

A few others came to Malaysia with the expectation that they would be able to earn a 'lot of money'. However, they claimed that this expectation remained unrealized. The fact was that if they went back home they would not be able to come back again. Therefore, they did preferred not to visit or return voluntarily to their homelands. Naturally, recouping the money spent on financing their migration was more important to workers than making pleasure trips. Obviously, the frequency of visiting home country is higher among the BRHKs compared to the BRMs. Financial factors occupy a significant place in rationalizing the migration decision and subsequent post-migration decisions about whether or not to make visits home. The following chapter analyzes diverse issues of financing migration.

The case of Akmal substantiates the lesser extent of visits by the BRMs than the BRHKs to Bangladesh. In an interview, Akmal, aged 36, said that he visited Bangladesh twice during his 11 years of residency in Malaysia. He visited Bangladesh for the first time when his contract with a company for five years expired. Later, his contract was renewed for two more years. He visited Bangladesh for the second time after six years of his first visit. It is worth mentioning that during his first visit to Bangladesh, the company (a paper mill in Selangor, Malaysia) provided him 50 percent of his airfare. However, for his second visit, he did not receive any support from his employer. During his visit to Bangladesh, he bought a few electronic items and other goods such as a camera, a blanket and a watch. He sold the camera and watch at the Stadium supermarket and the blanket at Gulistan Super Market, both of them are large markets in Dhaka, earning a good profit. Last year, he sent another camera through one of his friends who visited Bangladesh, which his wife sold (Source: Study interviews, Malaysia).

This section sheds light on the problems the migrants encounter overseas. Increasing number of migrant workers have been complaining of low pay, long hours of work, substandard dwelling and working places, and insufficient protection measures. These problems are further compounded by language problems and lack of knowledge on basic rights (BBC 2006). Therefore, problems are very significant issues in rationalizing the migration decision. The dynamics and the severity of the problems faced by the BRHKs and the BRMs vary widely. Hence lists of problems are presented separately. Data show that the BRHKs faced considerably less problems as compared to those faced by the BRMs. 'Unpredictable income level' was the most pronounced problem for the BRHKs followed by the 'long hours waiting on the street'. The lowest number of respondents stated the problem of 'hassle in visa extension'. A significantly higher number of the respondents mentioned about the 'cost on visa extension'.

> ... we have built Malaysia. The bricks are the witnesses of our labour we devoted. Our sweat and our labour mixed with the asphalt of the roads, concrete of the bridges and skyscrapers that made Malaysia known worldwide. Once done, we were dumped away. (Hashem, a migrant in Kuala Lumpur)

Like Mr. Hashem, many others expressed their grievances in the same way. Bribing police was the most pronounced problems faced by the BRMs. However, some other BRMs said bribing police is not a problem; rather they were able to prolong their stay in Malaysia by bribing. 'Lower salary than promised'[16] was the second on the list of problems, followed by 'insult'.

The factor analysis

As shown in the previous sections, a long list of problems emerged from the interview with both the BRHKs and the BRMs. This section attempts to reduce them into factors in order to explain the phenomena more clearly. Factor analysis was applied to reduce the list into three factors: employment; social factors; and networks. Employment factors have been grouped on the basis of problems related to employment, social factors on social problems faced, and networks factors on problems related to the respondents' network. The analysis shows that the employment factor explains 68.917 percent of total variance (Eigenvalue is 8.269) implying that the factor could reasonably be adequate in terms of explaining respondents' reasons for expression dissatisfaction and grievances. The

---

16   As Saleh (2007) was narrating of a potential migrant's story in which he sold his *vita mati* (all assets) for taka a lakh or two (US$1,500 or 2,500) to make his way to Malaysia to do slave labour, in order to change his fortune and break the invisible glass ceiling – only to be cheated by the *adam bepari* (recruiters) and return home penniless. If a migrant in such a situation is lucky to return alive, then the only remaining options require him to choose between becoming a professional *mastan* or a bomber for the cause of Allah. One might say that such a class struggle exists in every country, but striking in Bangladesh is that, increasingly, every passing year, the gap between what's offered for the rich and what's offered for the middle class is becoming wider, not to even mention the poor.

social factor is also significant, as it explains 69.313 percent of the total variance with Eigenvalue of 7.625. Clearly, employment related problems were the most significant in rationalizing migration decisions.

This section deals with the sources of assistance available to solve the problems of migrant workers, who require support at every stage of the migration process. In a new environment, they expect support mainly from the respective consulate office, friends and relatives abroad (Khalaf and Alkobaisi 1999). However, the need for help does not end as soon as they arrive at the destinations; rather different kinds of problems arise after their arrival.

This section highlights the source of help for the migrants who faced problems, sourced primarily from compatriots, employers, relatives and companions. Migrants normally use the ties of kinship and friendship to settle down in the new environment, to secure accommodation, to learn language for survival, and to adjust to local cultures. Most of them knew the whereabouts of their other compatriots and their contact points. They stayed in touch to give social support, especially, when any of their peers fell into trouble with the authorities or in terms of personal finance.

Assistance normally offered from the consular offices is different. In an interview with Hong Kong patrol police, they said that Bangladeshi passports are hand-written, not digitized, and therefore not machine readable. As a result, they are very difficult for the authorities to authenticate, leaving the migrants vulnerable to police interference. Hong Kong police often seek help in this regard from the Bangladeshi Consulate. Indeed, the consulate reported handling an average of three such cases a day. Another report says that in other countries in Europe between 30 and 40 migrants seek help from Bangladeshi Consulate or High Commissions per day (*Daily Ittefaq* 2007).

Chapter 7
The Dynamics of Income and Remittance

This chapter addresses the income level and the personal expenses of migrant workers, as well as their flow of remittances (in cash and kind), which is defined as the portion of earnings sent back from the country of employment to the country of origin (Russel 1986, Doorn 2000, Carling 2005). I will provide an overview of the mechanism of transfer of remittances, as well as its uses and impacts. This overview will depict how Bangladeshi migrants use formal and informal channels to send their remittances to Bangladesh. The available statistics on remittances often do not match the actual amounts that migrant workers report transferring to Bangladesh each year. I will provide an analytical framework which illustrates the preferred options for remitting money to Bangladesh. This further indicates that migrant workers' expected income while abroad is the primary driving force behind their migration decisions. As I have argued throughout the book, the savings that migrants can make from their income after deducting expenses for subsistence in the country of employment is seen as one of the principal factors by which they justify their migration decisions.

In recent years, the large amount of remittance that flows from overseas into the economy of Bangladesh has generated a substantial impact on the country's development. However, this contribution made by migrant workers has not been widely recognized until very recently. The main source of data on migrants' remittances is the annual balance of payment records which are compiled by the International Monetary Fund (IMF 2005, Russell 1992, Haas 2005). Global estimates of official remittance flows suggest that remittances increased from USD 43.3 billion in 1980 to USD 70 billion in 1995, and that over the past three decades the global amount has reached USD 100 billion per year (IMF 2005) and today it is around 300 billion (World Bank 2009). According to Gammeltoft (2002), around 60 percent of this remittance goes to developing countries. Thus, public policies to make use of migrants as a development resource will have to consider the amount of annual remittance and the roles played by migrants and their communities in the remittance processes (Gammeltoft 2002).

Like many other developing economies, remittances constitute the single largest source of foreign exchange in Bangladesh (IMF 2005). Remittances increased steadily until the early 1980s, reaching approximately USD 630 million in 1983 (BMET 2005). After a brief declining trend in the following year, the growth of remittance flows started again (IOM 2005, BMET 2005). This growth trend is illustrated by the fact that, while in 1976 only USD 24 million entered Bangladesh through official channels, the amount had topped USD 4,249.89 million by 2005 (IOM 2005, Bangladesh Bank 2006 and 2010). The steep fall after

2008–2009 does not indicate a decline in remittance flow because the estimation was incomplete (i.e., only including the first two months of the year).

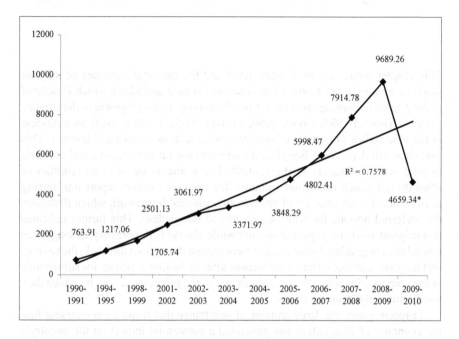

Figure 7.1 Trend in growth of remittances to Bangladesh (USD millions)
Note: * Data up to month of November of the financial year 2009–2010.
Source: Bangladesh Bank 2010 (February).

Between 1991 and 2005, USD 545.31 million was transferred to Bangladesh from Malaysia (BMET 2006), and between 1998 and 2007 USD 42.45 were received from Hong Kong in total (Bangladesh Bank 2007). The country-wise remittance flows show that Bangladesh earns most of its remittance from the Middle East countries. Between 1997 and 2004 (financial years), approximately USD 6,900 million was remitted from Saudi Arabia, more than USD 1,900 million from Kuwait, and more than USD 1,400 million from the UAE. The shares from Qatar, Bahrain and Oman amounted to several hundreds of millions in US dollars as well (BMET 2005).

The true value of remittances is likely to be much higher, as only a portion of total remittance flow follows official channels. Remittances conducted through informal channels are considerable in Bangladesh. It is estimated that the amount of such informal remittances would be at least double or even triple that of the recorded figures (Abella 1989, Puri and Ritzema 2004).

Table 7.1 Country-wise remittances earned (USD millions)

Country	2003–2004	2004–2005	2005–2006	2006–2007	2007–2008	2008–2009	2009–2010
Bahrain	61.11	67.18	61.29	79.96	138.20	157.43	56.35
Kuwait	361.24	406.80	454.38	680.70	863.73	970.75	331.74
Oman	118.53	131.32	153.00	196.47	220.64	290.06	115.31
Qatar	113.64	136.41	161.43	233.17	289.79	343.36	126.69
KSA	1386.03	1510.46	1562.21	1734.70	2324.23	2859.09	1089.89
UAE	373.46	442.24	512.64	804.84	1135.14	1754.92	607.24
Libya	0.13	0.27	0.16	2.61	0.36	1.25	0.84
Iran	0.38	0.52	1.68	2.36	3.24	3.28	1.47
Sub total	2414.52	2695.2	2906.79	3734.81	4975.33	6380.14	2329.53
Australia	4.79	7.15	8.89	11.34	13.11	6.78	3.01
Hong Kong	5.92	5.63	5.37	6.15	8.10	9.09	3.39
Italy	27.16	41.38	78.43	149.65	214.46	186.90	82.65
Malaysia	37.06	25.51	19.05	11.84	92.44	282.22	171.18
Singapore	32.37	47.69	61.32	80.24	130.11	165.13	63.19
UK	297.54	375.77	517.39	886.90	896.13	789.65	292.47
USA	467.81	557.31	701.37	930.33	1380.08	1575.22	489.75
Germany	12.12	10.10	10.95	14.91	26.87	19.32	7.18
Japan	18.73	15.99	8.71	10.17	16.29	14.12	4.96
S. Korea	5.19	18.41	16.40	17.08	19.69	18.33	8.64
Others	48.76	48.15	92.56	125.05	142.17	242.36	152.85
Sub total	957.45	1153.09	1520.44	2243.66	2939.45	3309.12	1279.27
Total	**3371.97**	**3848.29**	**4427.23**	**5978.47**	**7914.78**	**9689.26**	**3608.8**

Source: Bangladesh Bank 2010.

The above table shows the remittance flows from countries that constitute the major remittance providers for Bangladesh. As the figure shows, among the cases examined in this book – Hong Kong and Malaysia – Malaysia is a significantly higher source of remittances to Bangladesh. Remittances from Hong Kong, although not very high as compared to other major remittances sending countries, are much higher than those from many European countries and Australia. It is interesting to note that remittances from Hong Kong are even higher than those from South Korea where there exists a formal labour export agreement with Bangladesh.

The model for remittance flows

Post-migration income probability, which determines the 'wage differential' between the pre- and post-migration income, significantly contributes to making a migration decision. Therefore, income earning is a primary factor that migrant workers take into account before they make a decision to migrate. The cost-benefit equation for remittances is normally performed before the migration is undertaken (Siddiqui and Abrar 2003). While remittances can be sent in-kind, the term 'remittance' generally refers to cash transfers between or within countries. International remittances thus come from the earnings of migrant workers, whereas intra-national remittances are from the earnings of the internal migrants (Russel 1986, Miyan 2003). Intra-national remittances are not reflected in a country's Balance of Payments, however. In China, for example, an estimated 50 to 60 million internal migrants employed in the coastal areas remit a large part of their earnings to their families back home (Siddiqui and Abrar 2003), but this has no expression in official reporting.

In this section I will consider the issues of remittance from a broader theoretical framework to answer the question of what constitutes remittances and modes of remitting among Bangladeshi migrant workers. Modifying Russell's (1986) model on decisions to remit, I have devised an explanatory model for remittances (Figure 7.2).

The volume and scale of total remittance flows is determined by several factors, including the number of migrant workers present in a host country, wage rates, economic activities in both the host country and the sending country, exchange rates, political risk, modes and convenience for transferring funds, and marital status of migrants, as well as their level of education (Islam 2005). We have to consider whether or not the migrant is accompanied by dependants, the number of years passed since migration, household income level in sending country, relative interest rate between labour-sending and receiving countries (Doorn 2000). Black (2003) and Lucas and Stark (1985) identify three basic propositions that explain *why* remittances are sent back home. The first reason is the notion of altruism that migrants have to increase the well-being of their family in their home country by providing additional income. Secondly, the migrants intend to increase their assets and opportunities for their return; in this vein, the remittance also acts as a repayment for the past expenses of the family to finance the migration. The third reason for sending remittance often mentioned by the respondents is the pressure that the family exerts on the migrants. Carling (2005) underscores that the pattern of remittance is dependent on other characteristics, as well as the behaviour of the remittance providers and receivers. For example, the volume of migration in any given year determines, to some extent, how much the total income from remittances was. The IOM observes that remittances indicate how long the migrant workers have been staying in a given destination country (IOM 2003, ILO 2003, BMET 2005). Therefore the shorter the length of stay is the less is the income of the migrants. The model presented here demonstrates that remittances are carved out of the disposable income that migrants receive after deducting the cost of living expenses and, in some cases, savings and other costs. Additionally, migrants need time to settle in the host country before they can send remittances back to their own jurisdictions. Expenses matter in the cost analysis involved in the decision-making process.

Available pool of remittances

Migrant workers' wages can be rightly designated as the sole direct source of income for some receiving families. It is expected that income should be invested in a way that allows a profit to accrue. This fundamental concept of remittance as income is clearly set forth here. It is well established that business enterprises consider as income only profit or gain, not just gross revenue. All costs, expenses and allowable write-offs are first deducted from the gross revenue to determine the profit: this is the case for business enterprises. However, rules change for a wage earner. One must list every income source coming in and pay taxes on them without the benefit of the 'usual' business deductions. Taking these factors into account, the income benefits of the migrant overseas has been worked out below.

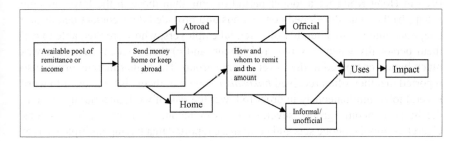

Figure 7.2 The remittance system model
Source: Modified from Russell 1986.

Data show that the mean income of the BRHKs was HK$8,580.36 (BTk. 72,933.06). The highest percentage (50 percent) of the BRHKs earned the lowest income range (HK$5,000–7000). Conversely, the lowest percentage of the BRHKs earned the highest range of income (HK$11,001–13,000). The mean income of the BRMs was RM 1175.[1] The statistical test shows a significant difference in income trajectories between the BRHKs and the BRMs (P=0.000). This means the BRHKs earned a significantly higher income as compared to the BRMs.

A question that needs to be answered is: if educational attainment and skills do not determine the level of wage of a labourer, why have the incomes of respondents in Hong Kong been so much higher than those in Malaysia? Most of the BRHKs state that they have simply been lucky. For example, if a group of seven or more workers are competing for a job and the employer only hires one of them, the chosen person would think that they have been lucky as there was an

---

1   To convert the currencies, rate on float in June 2005 was used. To work out the mean difference in income between the BRHKs and the BRMs, both the currencies (Hong Kong Dollars and Malaysian Ringgit) were converted into Bangladeshi Taka for analysis.

equal probability for any of the competitors to be chosen for the job. One migrant expressed this sentiment while pointing at the sky:

> ... everything depends on the wish of Allah (God). When an employer arrives with a work offer, we all approach him or her. S/he chooses one from among us. It is Allah (God) who knows who is going to be selected. (Rahman, a respondent in Hong Kong)

However, clearly income varied based on a number of factors other than perceived luck. My data show that in the case of the BRHKs, relatively newer migrants were likely to have fewer interpersonal connections and less competence, and thus were likely to get fewer job offers. The characteristics of the highest income group which allowed them to attain better job offers were the following: a) they had been in Hong Kong for a longer period of time than those in the lower income group; b) they had developed connections and provided their contact details and telephone number to a number of potential employers, who were then able to call them personally when they required labour; and c) their spoken Cantonese was much better, compared to the lower income group. The higher income group also reported that they often received *bakhshish*.[2] Some of them also managed to work several jobs simultaneously in order to lower the risk of work uncertainty. Among the highest income group, several had extra income sources, such as selling cigarettes, mobile phone sets and gold: these reported that having multiple sources of income was a superior way to achieve a higher level of income.[3]

As already demonstrated, remittances are drawn from the 'disposable income' that migrants receive after deducting the cost of living expenses and in some cases, savings and other costs. Directly after arrival, migrant workers usually need time to settle down, obtaining a living situation and some form of job before they are able to start sending remittances. The delay caused by expenses such as accommodation costs, which have to be met before sending remittance back home, is typical. The amount of remittance expenditure depends on the general expenditure pattern of the individual, determining how much they are able to remit. Respondents reported various patterns of expenditure, affecting the amount that they felt able to send home. However, my calculations show that, in general, the monthly expenditure

---

2   An amount or tip offered spontaneously on top of the fixed wage rate after work is finished when employers are pleased with the work done.

3   The income earners among the family members who remain in Bangladesh often cannot meet subsistence due to the limited income flow into the household economy from other sources. Whatever is the size of the family, all members tend to live in a shared economy, traditionally observed in Bangladeshi society. Generally, family members depend on a single income earner for family subsistence and if the family has a member overseas as a migrant worker, he or she is considered responsible to maintain the entire family. However, the migrants themselves also have personal expenses while abroad which must be taken from their earnings, increasing the pressure to obtain a higher wage.

of the BRHKs was higher than half of their monthly mean income, indicating that they usually remitted less than half of their total monthly income.

Table 7.2 Monthly incomes of the BRHKs and the BRMs

Income		
HKR (Hong Kong dollars)	**n=56**	**%**
5000–7000	28	50.0
7001–9000	6	10.71
9001–11000	16	28.58
11001–13000	6	10.71
Mean	8580.36	
SD	2064.366	
Total	*56*	*100.00*
MR (Malaysian Ringgit)	**n=70**	**%**
<500	4	5.7
501–1000	18	25.7
1001–1500	41	58.6
1501–2000	6	8.6
2501–3000	1	1.4
Mean	1175	
SD	385.869	
Total	*70*	*100.0*
Significance (BRHKs vs BRMs)	*P=0.000*	

Source: Author's field data, 2004–2006.

Data confirm that the majority of the BRHKs spent between HK$3,001 and 5,000 per month on personal expenditures and 29 percent spent between HK$5,001 and 7,000. For most of the BRHKs, accommodation cost constituted the primary drain on their income, followed by food expenses. Other costs included phone bills, entertainment, toiletries, and visa and travel fees (for entering China). Migrants who reported having girlfriends spent higher amounts, compared to those who did not. Mokles, a respondent in Hong Kong confided, pointing to one of his friends:

> He has girlfriends. He spends money lavishly on his girlfriends by giving gifts and watching movies – stupid indeed.

Around 44 percent of the BRMs spent in the range of RM 401–600 and about 43 percent spent between RM 200–400 per month. The mean expenditure[4] for the BRMs was RM 497.14 (Tk. 7954.24). As with the income difference between the BRHKs and the BRMs, there was also a significant difference in the expenditure amounts (P=0.000). In both samples, the amount of expenditure depended on the level of income earnings. Equally, the volume of remittances depended on the level of personal expenditure. It is clear that the lower level of income affected both the expenditure patterns and the remittance amounts of the BRMs.

Table 7.3 Monthly personal expenses

Expenditure		
BRHKs (Hong Kong Dollars)	**n=56**	**%**
1000–3000	3	5.36
3001–5000	34	60.71
5001–7000	16	28.57
7001–9000	3	5.36
Mean	5303.57	
SD	1393.748	
Total	*56*	*100.0*
BRMs (Malaysian Ringgit)	**n=70**	**%**
<200	1	1.43
201–400	30	42.86
401–600	31	44.29
601–800	7	10.0
801–1000	1	1.43
Mean	497.14	
SD	158.545	
Total	*70*	*100.0*
Significance (BRHK vs BRM)	*P=0.000*	

Source: Author's field data, 2004–2006.

---

4   To test the significance, Hong Kong Dollars and Malaysian Ringgit were converted into Bangladesh Taka (BDT). As of June 2005, when this study was being conducted, 1 HK$=8.50 BDT and 1RM=16 BDT.

Cash and kind

Various forms and modes of remittance transfers are in use worldwide. In many Asian countries, informal foreign exchange markets are fuelled principally by migrants' remittances. Two notable examples are the *hundi*[5] system used mostly by Bangladeshi, Pakistani and Indian overseas migrants (Kazi 1989, Menon 1988, Saith 1999 and 1989, Kardar 1992) and the 'Money Courier Industry' used in the Philippines (Alburo and Abella 1992). While cash is most common, multiple forms of remittances are in use worldwide; in Sudan for example, as pointed out by Brown in 1992, an estimated 80 percent of unrecorded remittances were transferred in cash while the remaining 20 percent was transferred in the form of smuggled goods (Puri and Ritzema 2004). However, remittances are commonly channelled into Bangladesh through two forms; one is in cash and the other is in kind.

Remittance in cash Cash transfer is the most used means of transferring remittances among Bangladeshi migrant workers. Data substantiated that all interviewed wage earners did not remit every month. Several reported that their income earning was not sufficient to remit funds each month after their own subsistence was met. Respondents normally transferred money every three to five months because to do so involves high transaction costs. However, the frequency of transfer depended on a number of additional factors: a) need for money by their families in Bangladesh; b) adequate savings; and c) availability of couriers to carry money. Couriers are frequent travellers who travel between destination and source countries for business purposes. They commonly supply t-shirts and clothes from Hong Kong and Malaysia to local retailers in Bangladesh.

The following table shows the statistics on remittances transferred to Bangladesh by the BRHKs and the BRMs. Analysis indicates that approximately 94 percent (out of 47 respondents) of the BRHKs transferred money in the month of the interview, while 71.43 percent (out of 50 respondents) of the BRMs did the same. The mean amount transferred to Bangladesh by the BRHKs was HK$3,196.43 (BDT 27,169.655), while for BRMs it was RM 475.0 (BDT 7,600), indicating that the BRHKs transferred a significantly higher mean amount (P=0.000) as compared to the BRMs.

---

5   The *hundi* system is the most important informal channel through which money is transferred by migrant workers back to Bangladesh. In this system the migrant gives money to an intermediary, who contacts an agent in Bangladesh who then bears the responsibility for dispatching the equivalent amount money to the recipient (an informal exchange rate is used). The recipient collects the money from the agent using a code that s/he receives from the migrant.

Table 7.4 Amount remitted to Bangladesh in the last month

Amount		
HKR (Hong Kong dollars)	**n=47**	**%**
1,000–3,000	18	38.30
3,001–5,000	13	27.66
5,001–7,000	13	27.66
7,001–9,000	3	6.38
Mean	3196.43	
SD	2461.852	
Total	*47*	*100.0*
MR (Malaysian Ringgit)	**n=50**	**(n=50)**
<300	4	8.0
301–600	14	28.0
601–900	28	56.0
901–1,200	3	6.0
1,201–1,500	1	2.0
Mean	475.0	
SD	347.845	
Total	*50*	*100.0*

Source: Author's field data, 2004–2006.

Remittance in kind   A number of studies show that transfers in kind constitute quite a significant portion of migrants' remittances (Ullah 2010). Carling (2005: 29) observed that the nature and volume of remittances in kind depends on the characteristics and the policy of the sending and receiving economies. Especially among the BRHKs, transferring remittances in kind has recently gained popularity due to a number of factors, namely ease of transfer process and price differential of goods between the sending and receiving countries increasing the popularity of this option. A number of the interviewed migrants preferred transferring in kind rather than in cash due to its inherent profitability in their eyes. One respondent expressed this view well: if he transferred 100 US dollars to Bangladesh, his family would get the BDT equivalent to 100 dollars. However, if he carried goods such as a second hand laptop or video camera (the demand for which is very high in Bangladesh), he would be able to sell the goods for an amount equivalent to 200 or 300 US dollars, effectively doubling or tripling the amount that he could transfer by carrying cash. Some items can be easily imported under the pretext that they are for personal use or are a gift (i.e. a computer or a camera). Migrants or their families are then able to sell these items in Bangladesh for a significant profit. Used laptop computers have become a popular item for transfer in kind, largely because they are a tax-free item and the demand in Bangladesh is high while prices in Hong Kong are relatively low.

Table 7.5 Volumes of remittances in kind

	Hong Kong			Malaysia	
Items	No of items	Market price+ (in Tk.)	Items	No of items	Market price (in Tk.)
Radio	14	7000	Computer*	–	–
Watches	23	23000	Radio	15	9000
Television	6	36000	Bags/briefcase	5	10000
Blanket	31	46500	VCP/VCR	3	21000
Cassette Player	12	48000	Cassette Player	7	28000
Others**	–	50000	Watches	21	29400
Gold***	–	100000	Others***	–	35000
Bags/briefcase	37	111000	Television	6	36000
VCP/VCR	26	156000	Video/DVD	2	40000
Video/DVD	17	425000	Blanket	28	42000
Camera	141	564000	Gold**	–	75000
Mobile phone	159	795000	Camera	40	160000
Computer*	97	1940000	Mobile phone	42	210000
Total	–	**4,301,500**		–	**695,400**

Note: '–' indicates not applicable; * Mostly used laptops; ** Clothes, shoes and home appliances for personal use; *** The exact weight or amount was not known; + Approximate.
Source: Author's field data, 2004–2006.

The volume of remittance in kind also depends on the frequency with which migrants are able to visit Bangladesh. As addressed in previous chapters, circumstances for visiting Bangladesh differ significantly between the BRHKs and the BRMs. The BRMs were more constrained due to high airfare, lack of money, limited holidays and, for most of the cases in this study, their legal status, impeding their ability to visit (and therefore to remit in kind). The BRHKs visited more frequently than the BRMs as they were more easily able to get their visas extended and passports issued or renewed. In addition, travel costs were not as high as for the BRMs. Thus, the degree of carrying goods as a form of remittance was higher among the BRHKs than the BRMs. Many BRHKs reported that Bangladesh Biman Airlines officers allow them to carry goods onto the plane even if their luggage is overweight, and at the check-out point in Zia International Airport, the customs and immigration officers would release them in return for a bribe. While visiting Bangladesh on holidays, the migrants brought or sent cassette players, video cassette recorders (VCRs), video cassette players (VCPs), computers, televisions, cameras, video cameras, clothes, blankets, home appliances, briefcases, radios, watches and gold, among other items (Siddiqui and Abrar 2003). Some migrants sought the help of friends visiting Bangladesh, asking them to carry items back home. Interview data revealed that not all of the respondents carried their items (remittance in kind) while visiting Bangladesh. However the trend is likely to change with the new immigration policy i.e. the new visa conditions imposed from 11 December 2006.

More than a dozen categories of items were transferred as kind, whether for commercial purpose or other purpose, with the most common being mobile phones and cameras. The following table indicates the range of goods sent by the migrants.

The mechanisms of transfers

While remittances in cash are transferred in a number of ways, they are categorized broadly into two: formal and informal. The latter has been mostly used by illegal migrant workers. Various circumstances make transfer through informal channels more attractive: Most of the study respondents were of the opinion that the slow process and higher charges of formal channels induced made them less useful than informal channels. Formal channels often call for legal documents concerning resident status, which leaves illegal migrants no other alternative than to choose one of the more popular informal channels, such as *hundi*. However, legal migrants also reported that informal channels have some advantages, such as a more rapid transfer, competitive rates of currency conversion[6] and manageable paper work.

Other reasons cited for the preference of informal channels over formal ones include the fact that illegal migrants have no access to the banking system in host countries. Siddiqui and Abrar (2003) and Ullah (2010) report other means of money transfer, such as, migrants sending visas to their families either in order to bring their family members abroad or for the family to sell to other potential migrants (Asia Migration Centre 2004). Puri and Ritzema (2004: 7) affirmed that migrant workers also tend to transfer money through an intermediary financial operator in the informal foreign exchange market. In this case, remittances are transferred by an agent outside Bangladesh via this agent's counterpart in Bangladesh to the

---

6 Transferring through formal channels requires the use of the government conversation rate, which is generally lower than the informal exchange market, reducing the attractiveness of this option: if money is transferred through informal channels, receivers can opt to sell their currency in the private market where the conversion rate is much higher than the official bank rate.

migrant's family or nominee in foreign currency that is exchanged at an agreed rate to local BDT.

According to the International Organization of Migration (IOM), transfers through informal channels account for 20 percent of the total remittances sent to Bangladesh (IOM 2005). However, research on remittance flows to Pakistan, Bangladesh, and Egypt indicates that informal remittances are estimated to be at least double or triple the recorded figures (Puri and Ritzema 2004). In Bangladesh's case, during the mid-1980s, the rate was as high as 80 percent (Figure 7.3). Although still high, Figure 7.3 illustrates the declining trend of unrecorded remittances. It is assumed that the recent introduction (in 2005) of Western Union in Bangladesh[7] – although on a limited scale thus far – has largely contributed to the precipitous drop in informal flows of remittances.

This study shows a different scenario from the current literature: In the case of the BRHKs, around 38 percent transferred their remittances through *hundi*, 27 percent through friends and compatriots, and 20 percent hand-carried it themselves. Although formal channels were available, in that money could be transferred from a bank in host country to a bank with a corresponding relationship in Bangladesh or through branches or subsidiaries of a Bangladeshi bank in the host country (Siddiqui and Abrar 2003), this study found that only seven percent of the total volume of transactions had been made through formal methods. This differs sharply from Skeldon (2003: 6) who found that 46 and 40 percent of remittance was sent through official and *hundi* systems respectively, while 4.6 percent through friends and 8 percent hand carried. However, this study's findings are supported by the findings that were found by Quibria though two decades ago. Quibria (1986, 1989), Merkle and Zimmermann (1992) and Quibria and Thant (1988) who argue that, due to lack of access to formal systems and the unfamiliarity with banking procedures in Bangladesh, unofficial channels are used more extensively.

Similarly to the BRHKs, this study found that 41 percent of the BRMs transferred money through *hundi* followed by friends and compatriots (22.86 percent), and only approximately 6 percent transferred money through official channels. The time duration required for transfer through *hundi* varied between 12 hours and two days. The demand draft has become by far one of the most popular official methods for remittance transfers as opposed to the telegraphic transfer, which has lost preference due to its time-consuming nature (IOM 2005). Sometimes officials, particularly in remote areas, do not disclose information to the clients in order to earn extra money. If the recipients pay extra, the process moves faster.

Due to direct links between banks or exchange houses in the destination country and Bangladesh, the transaction time should be shorter for formal transfers.

---

7 Western Union is a well-known international money transfer operator (MTO). In 2003, it had 182,000 agent locations worldwide (Akuei 2005, Carling 2005). Western Union has quickened the formal process of delivering money to the family members of migrants in addition to increasing reliability.

International remittances sent by bank draft to Bangladesh take on average 15 days. The available data on BRHKs and BRMs indicated only a marginal difference between usage of official channels and the *hundi* system. The following figure indicates that the BRHKs used couriers to a larger extent than the BRMs, possibly because couriers travel more frequently to Hong Kong than to Malaysia.

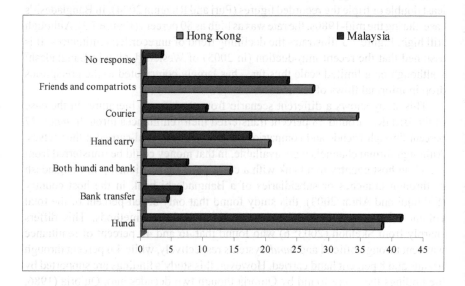

Figure 7.3 Channels used for remitting in the last year
(multiple response)
Source: Author's field data, 2004–2006.

Determinants of mechanisms

This section discusses why migrant workers chose a particular method of transfer.

Formal or official methods These are found to be the most secure methods. There are several common facilities for formal transfers including demand drafts, traveller's cheques, telegraphic transfers, postal orders, account transfers and ATM services. However, the study respondents suggested that a relatively longer transaction time has been one of the major problems in transferring money through official systems. The banking system reportedly does not respond as quickly as needed. A few respondents admitted that they had bribed the bank officials to speed up services. Moreover, there is evidence that a few of the respondents' lost significant amounts of their salary to thugs while coming out of a bank. Therefore, the formal system is not perceived to be risk-free. In addition, bank functionaries have reportedly harassed remittance recipients on trifling grounds or delayed payment on different pretexts in order to secure bribes (Siddiqui and Abrar 2003). Several respondents also reported that formal systems are difficult to understand for new users. Therefore many had to bring someone with them to help guide them through the formalities of various types of transfer.

Informal methods The respondents identified several grounds for preferring informal channels, despite acknowledging greater potential risks. Informal channels were often cheaper depending on the country of destination, and the exchange rates were higher than official transfer methods. Due to the limited use of paper work, in addition to its accessibility in remote areas, *hundi*, are easier to employ for a significant number of Bangladeshi migrants, particularly those not familiar with formal systems (IOM 2005). According to the respondents, informal methods offered more confidentiality than official channels – referring to concealing the occupation of the migrants and their reasons for remitting. According to most respondents, the foremost advantage of the informal system was the quick response. This combined with competitive rates, efficient and speedy private intermediation and the fact that *hundi* operators deliver money door to door increased the ease of the process for the beneficiaries. However, as Siddiqui and Abrar (2003) point out, a glaring problem associated with the *hundi* system is that, should the remitters loose money to the transfer agents, there is no scope for seeking legal redress.

In any case, despite possible preferences, illegal migrants, or temporary work permit holders have no access to a bank account in the host country, and are compelled to resort to informal transfer systems. The most crucial aspect for the *hundi* system, in order for it to continue to remain operational, are the additional services that it provides at the sending end as compared to the banking system, such as a better exchange rate, shorter transaction time and, in some cases, deferred payments (Siddiqui and Abrar 2003). Apart from this, some *hundi* agents also provide occasional individual services, such as sending letters or messages to families through the *hundi* agent. However, despite these benefits, most respondents reported that the safest method of transfer was by hand-carrying the money or goods themselves, although a few respondents reported experiencing hassles with airport immigration and customs.

A further push factor for using informal transfer systems are the high hidden fees charged by international money transfer agencies like Western Union and Money Gram, and the fact that these fees and exchange rates are not always communicated precisely to remitters (Siddiqui and Abrar 2003). However, despite this, few respondents remembered the amount they had been charged by banks or transfer agencies. According to information obtained from several banks on the transaction cost of remittances, the state banks Janata, Rupali, Agrani, and Sonali, each charge BDT 100.0 (around USD 1.5) as a service charge when they process a draft handed over to them by a remittance receiver at their local branches. Several study respondents who received money through the banks reported that they incurred additional costs, which Siddiqui and Abrar (2003: 39) termed 'speed money' (effectively a bribe) to facilitate and expedite the transaction.

This section discusses who the recipients of remittances are and how they spend or use the remittance. Focusing on individual migrants, a more detailed and heterogeneous picture of the dynamics of remittance usage emerges. Generally speaking, migrants' earnings may be consumed, invested or saved in the country of destination or may be saved in the home country for future consumption. As already observed, the proportion of the income that migrants remit depends on a number of factors. However, data from the study show that only a small portion of remittances are used for productive investments. Instead, the lion's share is normally used for debt payment, subsistence, house renovation or construction and family health care. Lindquist (1993) in his research found a similar trend of expenditure patterns for remittances. Carling (2005: 17) offers one explanation as to why the majority of the money goes to unproductive use: he cites family members as not being efficient in using the money; the absence of the remitters from the family increases the likelihood that the family members will not value the work required for the money and will spend lavishly.

As mentioned earlier, migration takes place mainly for its economic benefits. Therefore, remittance is an important variable by which migrants justify their migration decisions. Rationalization of the migration decision, as well as the end impact of the migrant's work investment, depends largely on how the remittances are used at home. A vast body of literature on migration is concerned with the division of remittance expenditures between investment and consumption. Broadly, the uses of remittances are divided into two: productive and non-productive. Over the years, concern has been expressed about the limited extent of the productive use of remittances. Although various authors have different perceptions concerning what constitute a *productive use*, there is a general agreement that the bulk of the remittance is used in daily expenditures on food and clothing. Some authors include land purchase and children's education as daily consumption; while others omit these, and include them in the category of productive uses (IOM 2005). This study has taken a conservative approach and defined daily consumption simply as expenses that are incurred in buying food and clothing.

Remittances also sustain intra-family obligations, despite the geographical separation of family members. Carling (2005) argues that in non-welfare countries like Bangladesh, it is taken for granted that adult children should support their parents, and migrants working overseas are expected to provide support financially by sending remittances. The household strategy theory suggests that family members also support the potential migrant in a number of ways. Other than monetary support, psychological support is essential, and is extended with a general expectation that the migrant would support the family at a later stage. However, family expectations of the migrant often run very high, as the family is often not aware of the situation in which many migrant labourers live and work. Many migrant workers are averse to letting their family know about the negative aspects or the miseries that they face abroad because this would add psychological stress to their families (IOM 2005).

The data from this study indicates that, in most families, it is the father of the migrant who received the money, followed by wives and mothers. According to the respondents, they remitted to support their parents, wives, children and sometimes their close relatives and friends. A few stated that they remitted to finance migration of other relatives. In terms of the number of persons supported by remittance, the data from the current study show that approximately 34 percent of the BRHKs sent remittance to support 7–9 members at home, while 44 percent supported 4–6 members. The mean number of family members supported by remittance from BRHKs was calculated to be 6.21, which implies that the dependency ratio is higher than the national level (4.7) (GoB 2009). About 55 percent of the BRMs supported 4–6 members at home and about 28 percent supported 7–9 members, resulting in an average of 5.34 family members, slightly less than the BRHKs but still higher than the national level.

This section identifies the uses of remittances, detailing where and what percentage of remittances are used for particular purposes. The pattern of usage of remittances often depends on the investment orientation of the respective families: there are some families who indeed want to invest in productive schemes, while others are inclined to display their new wealth and still others inclined to spend on education. In addition, the releasing of mortgaged land, leasing land, repayment of loans (borrowed for migration and other purposes), investment in business, savings and fixed deposits, insurance, social ceremonies, financing the pilgrimage to Mecca (*Hajj*) of relatives, financing migration of other family members, buying furniture and other home appliances were the major sources of spending.

Carling's (2005: 30) study indicates that only a very small portion of remittances were used for any productive investment. In many countries, surveys indicate that the majority of remittances are used for daily consumption (Siddiqui and Abrar 2001). Another study found that 50–60 percent of remittances in Asia goes toward consumption and only about 10 percent goes into investment. Education receives a small share of remittances, as does health care (5 percent) in Siddiqui and Abrar's (2001) study and around 4 percent in Murshid et al.'s (2002) survey.

In agreement with previous research, this study found that, for both the BRHKs and the BRMs, consumption was the most significant use of remittance. Thirty percent of the BRHKs and 41.5 percent of the BRMs spent the remittance for consumption. Between 6 and 7 percent of remittances were used for financing the education of siblings at home. Another important use of remittance for the BRMs is medical treatment and releasing mortgaged land and financing the migration of other family members. In Bangladesh, considering the resource scarcity, investing in land is the safest mode of savings. Arable land provides direct economic returns through crop production and the price of land usually increases over time (Mujeri and Khandker 2002). The use of remittances in releasing mortgaged land is also quite important in the rural context as it re-establishes the right of the person to cultivate their own land.

A relatively large portion of remittance expenditure (10 percent) is devoted to social ceremonies, such as weddings, child naming ceremonies, *khatna* (circumcision), Eid,[8] funerals, and funding dowries and the pilgrimage (*hajj*) for family members. Among these, the highest portion was spent on wedding ceremonies for family members and dowry, often paid when migrants' sisters were married off. Another important familial investment resulting from remittance is the financing of migration for other family members, involving 3.6 percent of the remittance sent by the BRHKs and 2.85 percent of that sent by the BRMs. For the families interviewed, this was treated as a very important investment for further enhancement of the familial income.

Table 7.6 Uses of remittances

Hong Kong			Malaysia		
Purposes	In Taka	%	Purposes	In Taka	%
Education*	975265	7.00	Homestead land purchase	–	–
Release of mortgaged land	104493	0.75	Education*	304351	6.15
Homestead land purchase	167188	1.20	Furniture	54437	1.1
Taking mortgagee of land	250782	1.80	Release of mortgaged land	59386	1.20
Repayment of loan	292579	2.1	Send relative for pilgrimage	89078	1.8
Furniture	334376	2.4	Gift/donation to relatives	103925	2.1
Send relative for pilgrimage	390106	2.8	Investment in business	123720	2.5
Home construction and renovation	434689	3.12	Others***	133618	2.7
–	–	–	Sending family member abroad	141041	2.85
Agricultural land purchase	487632	3.50	Repayment of loan	158362	3.20
Gift/donation to relatives	487632	3.4	Home construction and renovation	197952	4.00
Sending family member abroad	501565	3.6	Savings/fixed deposit	207850	4.2
Medical treatment	557294	4.0	Agricultural land purchase	272184	5.50
Others***	603271	.33	Social ceremonies**	336518	6.8
Investment in business	696618	5.0	Taking mortgagee of land	358788	7.15
Savings/fixed deposit	975265	7.0	Medical treatment	358788	7.25
Social ceremonies**	1393235	10.0	Consumption	2053752	41.50
Consumption	5294293	38.0	–	–	–
Total	**13,946,282**	**100.0**	**Total**	**4,953,749**	**100.0**

Note: '–' indicates not applicable; * For children and siblings; ** e.g., wedding, khatna (circumcision); *** Accidental and urgent need.
Source: Author's field data, 2004–2006.

---

8   The most important Muslim festival: Slaughtering cows or goats during Eid is a religious requirement and can constitute a significant point of spending.

For both respondent groups (BRHKs and BRMs), the amount of savings was relatively low when compared with other avenues of expenditures. For the BRMs, savings (in official banks) amounted to only 4.2 percent of the received remittance, while the BRHKs were slightly higher at 7 percent. Several respondents' families made fixed deposits and a few others reported purchasing insurance policies. The following table shows that despite the fact that the number of remitters from the BRM sample was higher than from the BRHK sample, the BRMs contributed less than half the amount compared to the BRHKs' remittances.

The development impact

This section discusses the impact of remittances on the well-being of the receiving family members and the national economy. Migration is one of the most important factors affecting economic relations between developed and developing countries in the twenty-first century, and generates a huge amount in remittances which have a profound impact on the living standards of the country of origin's population. Development processes can both affect and be affected by migration (Carling 2005, Sinclair 2001, Skeldon 1997, Blayo 1989), and, the effects of migration on countries of origin largely can be seen as either helpful or detrimental to development. There can be virtuous circles in which recruitment, remittances and returns (three Rs) speed up economic growth in countries of origin (IOM 2005), as well as vicious cycles, in which migration drains the work force and acts as a deterrent to development. In the case of Bangladesh, migration in search of employment has long played a crucial role in shaping the economy, not least by easing the unemployment crisis.

Some sources indicate that remittances are often double the size of official development assistances. The magnitude of remittances to Bangladesh exceeds the net earnings of the garment industry which is the highest foreign exchange generator and the largest employment creator in Bangladesh, amounting to more than USD 3.5 billion per year. However, Bowring suggests that the effective use of the remittance endowment for effective governance, so as to infuse it as a factor in furthering economic progress (see Bowring 2005). According to Brown (1995), Athukorala (1993), and Swamy (1981) the level of remittances constitutes very significant in proportion to a country's exports. For example, in Bangladesh, remittances were equivalent to about 44 percent of total merchandise exports in 1993; in India, about 13 percent in 1990; in the Philippines, about 22 percent in 1993; and in Pakistan, about 24 percent in 1993 (Puri and Ritzema 2004). In 2002, the flow of international remittances to developing countries stood at $80 billion per year, which is significantly higher than the total official aid flows to the developing world (World Bank 2003, Levitt and Nyborg 2004, Carling 2005).

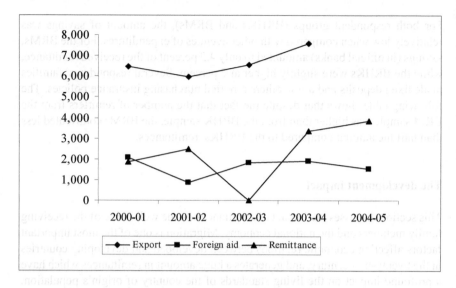

Figure 7.4 A comparison of income derived from export,
              foreign aid and remittance (USD millions)
Note: Data not available for exports after 2003–2004.
Source: Bangladesh Bank, economic data 2006.

Important to note is the fact that remittance has increased by 150 percent in the last
five years in Bangladesh and that it feeds the foreign exchange reserve and sustains
the value of local currency (BDT). Remittances inject capital into the economy
and, considering the fact that most migrant workers are unskilled or semi-skilled,
the majority of the funds coming in effectively reach out the poorer strata of the
society. The World Bank Report 2006 estimated that remittances reduced poverty
levels by 6 percent in Bangladesh (Blanchet 2007), which is of course near-
contrasting claim given the fact that the percentage of population living below
poverty line has reduced by around two percent in the last two decades (GoB
2008). It is often argued that remittances play a stabilizing role, conspicuously the
case during the Asian financial catastrophe of the 1990s, during which remittance
flows supported household expenditures and offset the precipitous decline in capital
inflow. International migration and hefty remittance flows have been prominent
features of the Bangladeshi economy for several decades. In particular, during the
1998–1999 financial years in Bangladesh, 22 percent of the total national debt for
imports was paid for by remittances. In 1977–1978 financial years, remittances
contributed 1.0 percent to the GDP of the country, rising to 5.2 percent by the
1982–1983 financial year, with a slight (4 percent) decline in the following decade
(Rushd 2004, IOM 2005), then reaching 6 percent in 2005 (Karim 2006). However,
this is still relatively low, when compared to countries such as Nepal which had a
15–20 percent contribution of remittances to its GDP in 2001 (Seddon, Adhikari

and Gurung 2002). According to the IOM (2005) in 1999, Bangladesh was ranked sixth in the world in as a remittance receiving country although in absolute figures remittances to Bangladesh represent only two percent of the global remittance flow (IOM 2005). According to the data published by the BMET, Bangladesh received a total of USD 31,683.85 million (BDT 145,353.96 crore) in remittances between 1976 and 2004 (BMET 2005 and 2006).[9]

The data provided by the Central Bank of Bangladesh closely supports Skeldon's (2003: 1) statement that the money sent back by migrants contributes more to national and local economies than trade does in several parts of the developing world and is substantially greater than the flows of development assistance funding in many countries.

Chami and others (2003) argue that where remittances are motivated by altruism, they tend to have a counter-cyclical impact, as family members receive increased remittances during economic downturns. To the extent that remittance earnings reduce the recipients' need to work, this may actually have a negative impact on overall economic activity. However, the impact of remittances upon the economy of the countries of origin depends largely on the way in which they are used (Miyan 2003). Remittances of overseas Filipinos are expected to reach US$14.7 billion in 2007, up $1.9 billion from 2006. In many low-income countries, remittances represent a significant percentage of the gross domestic product (e.g., 26.5 percent in Lesotho; 16.2 percent in Nicaragua; 5.8 percent in Burkina Faso). In Sri Lanka, remittances surpass earnings from tea export and exceed income gained through tourism in Morocco. A significant rise in the contribution by remittances to the GDP of Nepal from 11.5 percent in 2000–2001 to 16.8 percent in 2005–2006 is a clear testament that its role in the national economy is gaining stature. It has now positioned itself as the top contributor to foreign exchange earnings for the last five years. The share of remittances in total foreign exchange earnings increased from 36.6 percent in 2003–2004 to 46.7 percent in 2005–2006, whereas the corresponding share of exports dropped to 29.4 percent in 2005–2006 from 34.5 percent in 2003–2004.

As stated earlier, the economic impact of remittances depends, in part, on the propensity of recipient households to consume or to invest. Remittances that are invested in productive activities contribute directly to the overall economic growth of the country. However, even remittances that are consumed generate positive multiplier effects on the larger economy in general (Blanchet 2002, 2007). This study considers the opinion of respondents on the overall development and well-being of receiving families, as well as their housing conditions. However, the table shows that the well-being of migrant-sending families significantly increased after migration (and before first remittance receipt, when judged against the three 'perceived variables' of well-being ($P=0.000$). However, respondents did not seem to think that this improvement was completely due to the contribution made possible

---

9 A possible reason for the increase in remittances might be that during this period the number of migrants from Bangladesh to overseas locations has been augmented.

by migration. Most of the remitters generally want the remittances they send home to be used towards income generating and productive investment purposes and to stimulate the well-being of the household (Vete 1995, Connell and Brown 1995, Ahlburg 1995, Rivera 1984). To understand the diverse motivating forces behind remitting, and the impact on sending family well-being, the following indicators were considered.

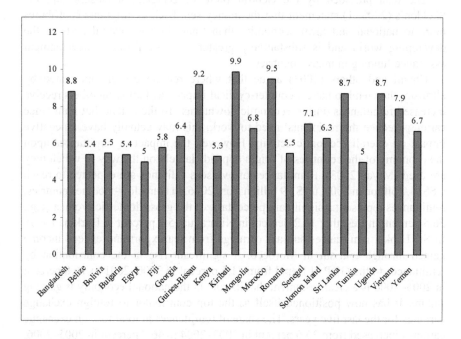

**Figure 7.5 Countries with remittances contributing to GDP
            from 5–10 percent**
Source: World Bank 2007.

To the respondents, housing conditions constitute one of the best indicators to measure the impact of any development that households go through as they derive a larger share of income from abroad (Blanchet 2007). Family Income and Expenditure Survey (FIES 2000) data show that a significant proportion of income from abroad, as a share of total household income, is spent on house renovation. This might seem irrational since migration is largely a lower and middle class phenomenon in Bangladesh, leading one to expect that income from abroad would be used for more immediate needs. However, renovating one's house is considered a status symbol in Bangladesh and as indicated in Chapter 5, an important part of labour migration is the matter of increased prestige. The current study data show that a considerable amount of remittance (BRHK 3.12 percent and BRM

4.0 percent) was spent on house construction and renovation. The following data clearly show that housing conditions have improved in the post-migration period for both BRHKs and BRMs households. The number of *kutcha* house has sizably decreased and the number of *pucca* house has increased significantly, indicating positive change (explanations of Bengali terms are the bottom of the table).

Table 7.7 Changes in well-being after migration*

Well-being status	Before	After	Significance
BRHKs	**n=56**	**n=52**	
Better than most of neighbours/better than average	–	3 (5.4)	
About average	21 (37.5)	42 (75.0)	P=0.000
Worse off, poorer than most of neighbours	35 (62.5)	7 (12.5)	
Total	56 (100.0)	52 (100.0)	
BRMs	**n=70**	**n=70**	
Better than most of neighbours/better than average	7 (10)	35 (50)	
About average	53 (75.7)	25 (30)	P=0.000
Worse off, poorer than most of my neighbours	10 (14.3)	10 (14.3)	
Total	*70 (100.0)*	*70 (100.0)*	

Note: '–' indicates not applicable; * In considering the well-being, land asset, ornaments, savings and other assets like cattle, etc., and food behaviour, education, living standard, health situation and health-seeking behaviour, asset building, savings, investment, and power (political and social) have been included; Significance at 95 percent confidence interval; Parentheses indicate percentages.
Source: Author's field data, 2004–2006.

As previously discussed, despite the fact that migratory processes are multidimensional and may generate a wide array of positive as well as negative consequences for development, remittances have lately become a demonstrative tool to measure the ties connecting migrants with their societies of origin. However, the dichotomous meaning of migrants' remittance, encompassing both monetary and social remittances, is relatively new, emerging only since the late 1990s (Levitt and Sorensen 2004, Lariosa 2006). While monetary remittance is defined as that portion of migrant's earnings sent from the country of destination to that of origin, social remittances are the ideas, behaviours, identities and social capital that flow from migrant-receiving to migrant-sending communities (Levitt 1998). Social remittances have been conceived of as resources and cultural flows that play a crucial role in 'promoting community and family formation and political integration'. To describe social remittances, they showed that migrants adapt knowledge of health awareness, cleanliness, orderliness and so on, to institute back in their home culture. However, the extent of such transnational social dissemination is very much dependent on migrants' own cultural specifications. This study has not put an emphasis on social remittances but has incorporated social remittances into the discussion of local adaptations (see Chapter 6). However, 'remittance equals to cash' theory has overshadowed other forms of remittances that only in recent times occupied reasonable space in literature. Social remittance with tremendous transformative significance can modify the economy, values, and everyday lives of a nation. Not all social remittances are positive, however. Even though social remittance is in the agenda, only the micro level impact is considered. It could however be argued that major development paradigm shift that took place over the decades in major economies in Asia has largely been the contribution of social remittances brought and sent to by, for instance, Gandhi in India, Ho Chi Minh in Vietnam, Shinawatra in Thailand, Mahathir in Malaysia, Suu Kyi in Myanmar.

Table 7.8 Changes in housing

Housing	Hong Kong			Malaysia		
	Pre-migration	Post-migration	% of change	Pre-migration	Post-migration	% of change
Pucca*	6	12	+100	2	5	+150
Semi-Pucca**	7	17	+143	9	12	+33
Tin***	10	14	+40	24	32	+33
Semi-Kutcha+	17	7	-69	16	17	+6.25
Kutcha++	16	6	-63	19	4	-79

Note: * Roof, fence and floor are cemented (with bricks); ** Fence is cemented (with bricks) but roof may be of tin and the floor may be of mud or cement; *** Roof is of tin, fence is of tin and the structure (i.e. pillars etc.) is of wood. The floor may be of mud or cement; + Either roof or fence is of tin or of straw and the floor is of mud; ++ Roof and fence are of straw with a wooden structure and the floor is of mud.
Source: Author's field data, 2004–2006.

Individual cost and benefit   At the individual level, migration involves costs and is expected to result in benefits to the migrants and their families. Migration is overall beneficial to an individual when the benefits gained substantially outweigh the costs invested (Miyan 2003). We need to have an idea of the gaps between the costs incurred and the remittances sent. This study shows that a Bangladeshi worker in Hong Kong spends a mean amount of BDT 78,267.86 (USD 1,151) for migration, while a Bangladeshi worker in Malaysia has spent a mean amount of BDT 109,142.86 (USD 1,605), most of which involves charges for an agent. The economic benefits of migration come in the form of direct remittances to the migrants' households. From the data gathered, it can be seen that the BRHKs remitted a mean amount of HK$3,196.43 (BDT 28,764 or USD 423) and the BRMs remitted an average of RM 475 (BDT 8,075 or USD 119) in the period of November–December 2005. This is apart from the amount retained by the migrant for personal use or to hand-carried during periodic visits or final returns home. The central personal benefits of working abroad are reported to be the higher income earned and the use of remittances for the improvement of living conditions, secured future and access to civic amenities for the migrants' families.

A few of the migrants (mostly the BRMs) said that their families could barely meet their subsistence in the pre-migration period, while their migration has enabled their families to increase to a level of household surplus. However, in most cases, the migration process for the BRMs completely exhausted their resources, and left their families destitute due to the loss of productive assets caused by their departure, which they were unable to make up for with their remittances. A notable opportunity cost that the migrants' families often pay is that migration reduces women's economic activities by enhancing their burden of household work, due to the absence of male members. This leaves the family even more dependent on the remittances sent by oversees members. In addition, the consequence of unaccompanied migration are family problems, often related to conjugal relations and demographic impact e.g., fertility and reproduction. The socio-economic implications of labour migration include the following:

Family economy The nature of migrant family labour has undergone many changes due to migration itself. Conventionally, the family unit tended to depend upon its members in creating their own self-employment for their collective economic survival, usually based on subsistence agriculture. At a basic level, family farms would draw sustenance from the land by raising animals and cultivating crops for commerce and self-subsistence, making clothes and building shelters. However, despite the monetary benefits of migration, labour shortages cannot be ignored: the outflow of labourers ultimately affects the family economy adversely. According to my respondents, the greatest negative impact is felt when they start sending money back to their homes, causing the other family members' economic productivity decline substantially i.e., the family members left behind become dependent on the remittances, often reducing their own level of work which leaves them more financially vulnerable than they were pre-migration.

Social costs and benefits Social costs involve both costs borne by the workers and their families, and those borne by the society at large. Notwithstanding that remittances do have a direct positive impact on the real income of both migrant workers and their remittance recipients (Siddiqui 2003, Miyan 2003), out-migration does impose significant social costs upon the societies that migrants leave behind, including the loss of talent and skills.

These costs and benefits have indirect impacts through multiplier effects. In terms of benefits, these can be extended to non-migrating households. For example, the migrant workers usually travel by their home country's airline, although this may depend on the travel agents or the recruiters. In addition, the home-country government receives income from workers' passport issuance and renewal fees (IOM 2005). Thus labour exports to foreign countries can generate revenue for the home country. Remittances, sent by the emigrating workers, often encourage consumption and inflation. According to existing findings, a significant portion of remittances are spent on consumer goods and house renovation (Miyan 2003). However, the possibility that remittances are used primarily for the consumption of imported consumer goods (World Bank 2009) has increased concern among policy makers regarding the extent of the impact remittances may have on domestic investment and thus the longer-term economic sustainability of local economies. Despite this, studies have demonstrated that higher levels of consumption are a major welfare gain and contribute to meeting basic needs that are otherwise sometimes poorly satisfied. Contemporary literature further highlights consumption as a productive use of human development (Rahman 2004). However, according to the point of view of most of the study respondents, that consumption is a waste. The respondents were less concerned with whether their siblings were obtaining sufficient nutrition; in their perspective the money they sent home was meant for investment in long-term productive uses.

Macro-economic cost and benefits   It is clear that Bangladesh receives remittances at the cost of the possible development of skill and expertise. Some researchers (for example, Siddiqui 2003, Rahman 2003) regard the use of remittances for consumption not as a waste, but rather as an investment in human development. However, from a micro-level perspective and macro level as well, the monetary cost involved in the migration process imposes some negative impacts on home country. While remittances have a positive impact, it is also argued that the inflow of remittances might cause inflation in the recipient country's economy. It also may aggravate income inequalities (Lowell and Kemper 2004). Despite this, the perception of late has been that the overall impact of remittances on the recipient economy is a positive one.

There is no doubt that the flow of remittances contributes to the improved living conditions of the recipients as well as improved income distribution in favour of the poorer and less skilled workers. Therefore, migrant households are marked by better human resources development than ordinary households (Miyan 2003). In Bangladesh, the increased remittance turns out to be the primary reliever of the acute foreign exchange shortage facing the country. This, in turn, has an impact on gross domestic product (GDP), annual development budget (ADB), foreign exchange earnings, merchandise export receipts, import payments, trade balances, current account balances and foreign aid in Bangladesh (Miyan 2003, Bangladesh Bank 2006), resulting in a significant impact at the macro level.

The preceding sections have analyzed the diverse causes of frustrations and problems faced by Bangladeshi migrant workers abroad: This section analyzes the level of satisfaction experienced, particularly in relation to aspects of income benefits and living abroad. While personal satisfaction partly reflects objective conditions, it is evaluated in the context of an individual's own subjective standards, increasing the complexity of the analysis. In this study, level of satisfaction has been assessed with the question: 'are you very satisfied, moderately satisfied, not satisfied or highly dissatisfied with you decision to migrate, with regards to your income?' Values for each response category were assigned accordingly.

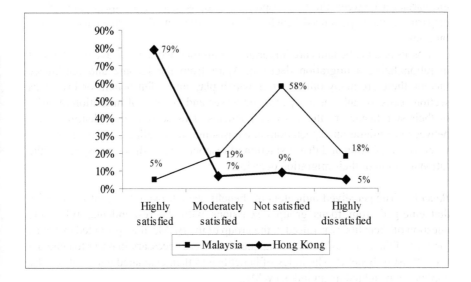

Figure 7.6 Comparative levels of satisfaction with income between BRHKs and BRMs

As addressed previously, income is the primary motivational factor for migrants to relocate abroad for work, hence level of satisfaction pegged to income benefits merits investigation. The preceding chapter (6) evaluated the various levels of income earnings: This section evaluates the level of satisfaction and dissatisfaction of the migrant workers with their income. Further, this section explores the well-being of the migrants based on the subjective[10] assessment of their own income.

The study data indicate that an overwhelming majority of the BRHKs (79 percent) are highly satisfied with the level of their income. Hence, as illustrated in Figure 7.5, the curve starts from the peak at the 'highly satisfied' point with

---

10   Subjective measures often tap into a number of economic and psychological considerations, ranging from consumption levels, perceived economic needs, and past and expected levels of economic well-being. People experience fluctuating subjective well-being depending on whether they experience pleasant or unpleasant emotions, pleasure or pain, and when they are satisfied or dissatisfied with their lives and decisions (see Chan et al. 2002).

steep fall toward the direction of 'highly dissatisfied': 'moderately satisfied' was endorsed by only 7 percent, not satisfied by 9 percent, and highly dissatisfied by 5 percent. Almost the opposite trend is evident in the income satisfaction level of the BRMs, with 58 percent endorsing the 'not satisfied' response, 19 percent the 'moderately satisfied' response, 18 percent the 'highly dissatisfied' response, and only 5 percent the 'highly satisfied' response. This finding clearly highlights the variation between the two respondent groups, and raises speculation about disparity in the experienced benefits, both for migrants themselves and for their families.

Losses and the benefits are important for migrant workers to take into account in rationalizing a migration decision. Apart from the income and remittance factors, there are many other factors which play into a felt benefit or loss. This section seeks to rank various perceived losses and benefits of migration in order of their significance to the study respondents. This section also identifies gaps between pre-migration expectations and post-migration realities, and measures the levels of satisfaction and dissatisfaction of respondents on different aspects of the rationalization of their migration decisions.

Benefits   The perceived indicators of 'benefit' were broken down into categories that emerged from focus group discussions with the respondents. When the question of 'benefits' was raised in the group of the BRMs, they provided instead a long list of their 'losses' and probing questions were necessary in order to generate a small list of benefits. The index of benefits was then evaluated and analyzed by applying a weighted mean index (WMI).

The most significant indicator for respondents was that most of them were unemployed in Bangladesh, but were able to secure jobs in the host countries. This was the case irrespective of their satisfaction with their income earnings. Thus, this study indicates that the perceived indicator 'got employed' received the highest endorsement with a WMI of 0.778 for the BRHKs and 0.556 for the BRMs. This was followed by 'ensured higher income' which ranked second with 0.763 WMI. These findings again reconfirm the migrants' unemployed status in their home country as analyzed in their profile (Chapter 2). Most respondents reported that they at least in the host countries they have some work to do. The indicator 'future is secured' ranked third with 0.704 WMI, followed by 'increased assets' with 0.701 WMI. 'Constructed or renovated house' ranked 10th with 0.604 WMI, followed by 'ensured children's education' with 0.606 WMI. These findings correspond to the analysis of remittance use as discussed in the preceding chapter (6).[11]

---

11   Perceived benefits expressed by BRHKs are ordered according the WMI: Got employed = WMI 0.778; Ensured higher income = WMI 0.763; Future is secured = WMI 0.704; Increased property/asset = WMI 0.701; Escaped evil persons = WMI 0.672; Earned esteem in the locality = WMI 0.670; Repaid loan = WMI 0.670; Earned overseas experience = WMI 0.654; Ensured better life = WMI 0.638; Ensured children's education = WMI

Losses This section assesses the perceived losses of the migrant workers, also measured by weighted mean index (WMI). The question asked was: 'What did you lose through migration?' Importantly, the study respondents perceived that, due to the length of separation from their families, they had 'lost' their family ties. The BRMs ranked the loss of family ties as their primary felt-loss with a 0.790 WMI. This was followed by 'loss of assets' with a WMI of 0.683: Most of the respondents had to sell their assets to finance their migration and yet many of them could not buy these assets back (*Daily Ittefaq* 2007). The BRHKs ranked the slightly different 'loss of trust in the family' as their primary loss with a 0.112 WMI, followed by 'loss of health' with a WMI of 0.106. According to the respondents, trust denotes a feeling of certainty of commitment (especially moral) made by a person, and in this case, loss of trust may be related to financial or marital fidelity. This is further indicated by the BRHKs ranking of the 'loss of family ties' fifth, with 0.088 WMI, followed by 'loss of sexual morality' with 0.086 WMI. 'Loss of youth' ranked third for the BRMs, with 0.668 WMI. Most of them felt that they had lost the prime of their life, as indicated by the quote below:

> We have wasted our lives. We have lost our future, family and everything. We have lost our dignity both here and in our country. (Amir, a migrant in Klang, Malaysia)

Difference in the perception about losses between the BRHKs and the BRMs is obvious from the WMI values. BRHKs: Loss of trust in the family = WMI 0.112; Loss of health = WMI 0.106; Loss of mental peace =WMI0.102; Loss of ambition = WMI 0.095; Loss of family ties = WMI 0.088; Loss of sexual morality = WMI 0.086; Loss of property/assets = WMI 0.083

BRMs: Loss of family bonds = WMI 0.790; Loss of property/assets = WMI 0.683; Loss of youth = WMI 0.668; Loss of enthusiasm = WMI 0.667; Loss of ambition = WMI 0.652; Loss of mental peace = WMI 0.636; Loss of trust in the family = WMI 0.610; Loss of control of landed property = WMI 0.609; Loss of sexual morality* = WMI 0.513; Loss of roots = WMI 0.479; Loss of health = WMI 0.427; Loss of caliber = WMI 0.402

Where do returnees end up?

The preceding chapters have dealt with aspects of pre and post-migration periods. In dealing with these issues, diverse bad and good experiences; satisfaction and

---

0.606; Constructed/renovated house = WMI 0.604. Perceived benefits expressed by BRMs: Got employed = WMI 0.556; Escaped evil persons = WMI 0.121; Repaid loan = WMI 0.114; Constructed/renovated house = WMI 0.114; Earned overseas experience = WMI 0.096.

dissatisfaction related to a number of factors of migration have been analyzed. This section deals with the issue of return migration. This last phase of the migration process involves the return and reintegration into the home country. The theoretical literature on migration treats return migration as a part of life-cycle planning where return migration is an optimal decision-making phenomenon (Djajic and Milbourne 1988, Dustmann 1997, Stark, Helmenstein and Yegorov 1997). In terms of the study population, it has already been addressed that there is a gap between their pre-migration expectations and post-migration realization and that this gap is even wider in the case of the BRMs. Therefore, in order to know how they want to start their life again after they return, they were asked about their future plans, after returning home.

The study shows that majority of the BRMs (59 percent) were planning to start a business after returning to Bangladesh, which supports the findings of Abdul-Aziz (2001: 9). Significantly, many of the remainder (approximately 39 percent) had not planned for after their return. The respondents reiterated that they felt the prime period of their life had been spent in the Malaysian factories or palm and coconut plantations. Many of the BRMs expressed an intention to stay in Malaysia permanently, hoping to develop roots there, some were even determined to marry to Malaysian women. However, marriage to a Malaysian does not guarantee permanent residence and many of these efforts went in vain, resulting in divorces or forced separations in the end. The majority of the BRHKs reported that they had no idea also what to do after they return, likely because these respondents intend to stay in Hong Kong as long as they have work, and many of them are trying to use Hong Kong as a gateway to migrate to other countries. Therefore, returning home is not their immediate or preferred plan, and not something to which they give much consideration.

As evidenced by the case studies in this book, migration in order to take advantage of options in the overseas labour market was perceived as an alternative to escape local limitations (Lindquist 1993), poverty and political persecution. Other reasons cited are in order to have an easy life in old age, to support the family and to acquire sufficient capital to start a business. Evidently, many emigrants leave their countries with an intention to return, but often they do not. All of the cases in this study offered unique information relating to the migration process, trends, factors and wage related issues, as well as the role of agents and broker networks. Deception and false promises on the part of agents has become so obvious that it is now considered a given fact and difficulty in the process. To arrange money for financing migration, almost all of the returnees said that they had had to sell their valuables, including land, agricultural products and jewellery. The primary decision 'to go' or 'not to go' has is usually made collectively and, in most cases, at the family level.

Diverse reactions were expressed by respondents on their migration experiences. For example, a few of them would like to re-migrate, while others oppose the idea, saying that if the income overseas is not significantly higher than the probable income in home countries there is no point in migration. They believed that they could earn the same amount of money at home with the same amount of labour. However, with very few exceptions, in the case of returned migrants from the BRHK sample, none reported being able to start a business with the money they had saved from their work abroad, because it was not sufficient. In general, Black (2003) states that: remittances are likely to be higher in situations where the migrant has left for broadly economic reasons rather than political or social reasons, where they have temporary rather than permanent resident status, and where they are young, but married with family left behind at home. Remittances will increase as emigrant wages increase although, at a certain point, this trend levels off and increases in wage levels no longer translate into higher remittances (Black 2003).

Return migration has been seen as a key link between migration and development. As cited in Carling (2005), emigrants can be crucial resources in the development process of their countries of origin *if* they return with capital, experience and relevant skills. However, when return migration is hindered by complications with the migration process, perhaps leaving families in greater debt than prior to migration, due to fraudulent agents and brokers, or when the consequences of migration – both bad and good – reduce the likelihood of a healthy return, migrants lose their ability to contribute meaningfully.

Table 7.9 Summary table of case studies

Cases	Age	Education (years of schooling)	Marital status*	Migration experience (years)	Received help for migration from	Overall opinion on migration**
Case 1	27	12	M	6	Relative/Agent	MS
Case 2	32	10	S	5	Relative/Agent	DS
Case 3	30	14	S	5	Relative/Agent	MS
Case 4	37	10	M	7	Relative	SA
Case 5	37	10	M	5	Relative/Agent	DS
Case 6	39	8	M	5	Relative/Agent	DS
Case 7	36	14	S	7	Relative/Agent	MS
Case 8	35	12	M	7	Relative	SA
Case 9	33	10	M	10	Relative/Agent	DS
Case 10	30	10	S	6	Relative/Agent	MS
Case 11	34	10	S	1	Relative/Agent	MS
Case 12	40	12	M	5	Relative	MS
Case 13	38	8	S	6	Relative/Agent	DS

Note: * M: married; S: Single. ** SA: satisfied; MS: Moderately satisfied; DS: dissatisfied.
Source: Author's field data, 2004–2006.

Chapter 8
Rationalization and the Implications

The rationalization of the migration decision before, during and after migration by migrant workers has been central to the current study's research question. Existing researches have only sporadically and in a cursory manner touched upon issues of decision-making and rationalization. Therefore, bringing the two factors under study with the objectives of understanding the components of pre-departure decision-making on the part of the migrants and their families, and then the rationalization of their decisions post-migration, both in situations where the migration goals have been achieved and in cases in which they have not, fills significant gaps in the migration literature for this region.

This concluding chapter synthesizes the previous chapters and analyzes why migrants may rationalize their migration decisions in particular ways. In order to set rationalization in an appropriate context, the migration process was analyzed from two perspectives: pre- and post-migration. Such rationalization is a repetitive process whereby migrants begin 'explaining' their decision from the inception of the idea, and in this way, often unconsciously, re-frame the circumstances for themselves in ways that justify the hardships they have experienced as a result of their decision to migrate.

The key argument of this research is that Bangladeshi potential migrant workers from diverse socio-economic backgrounds decided to migrate overseas with a set of expectations in mind. However, in most cases, their expectations have remained unattained and they respond by rationalizing their migration decisions. This rationalization process enables migrants to come to terms with their unrealized expectations. For example, many of the respondents in Hong Kong confessed that Hong Kong was not their target country of destination; however, it became so over the course of time. Many Bangladeshi migrant workers who have been working in Hong Kong had originally intended to only use it as a gateway to sneak into other countries, but having failed to do so and after becoming tired of waiting for the opportunity, they eventually rationalized Hong Kong as a conducive place to live and work. Although Bangladeshi migrants in Malaysia have experienced more difficulties than those in Hong Kong, even they are able to ameliorate themselves to their situations.

The notion of rationalization refers to justifying decisions that have been made. Rationalization is an ambiguous and controversial concept because it often involves assumptions (Elwell 1999). It is difficult to measure rationalization quantitatively, and it is very much actor-centred. Therefore, analyzing rationalization is indeed a challenging task because, in addition to its subjectivity, many unpredictable factors are embedded in both the construct and the context. However, the conceptualization of rationalization in my study, view of Habermas has been employed, wherein rationalization is a process involving the practical application of knowledge to achieve a desired end by observation, experience and reason in order to gain a sense of mastery over the natural and social environment (Elwell 1999). To the migrants, rationalizations might oscillate occasionally because this process is related to psychological states which are highly prone to fluctuations. One might include a particular variable in explaining rationalization while another might exclude it. In my study, for example, most respondents in Hong Kong justified their migration decision by the fact that they earn 10 to 15 times more than was possible in their own country, while most respondents in Malaysia rationalized their stay with the hope of recouping the costs they had incurred in financing their migration.

This section shows how the particular theories used in this research explain the migration contexts of rationalization. As already explained, economic factors, by far, are the most influential, both in terms of the migration decision and in terms of post-migration experiences. I argue that the cost and benefit model is not able to explain many of the qualitative aspects of the migration decision. The major flaw in the cost and benefit model is that it measures only objectively known quantitative aspects of a phenomenon. This means that it does not take into account people's expectations and assumptions, which cannot be easily quantified – e.g., the expectation of a higher income, with no amount specified, or migration viewed as a matter of prestige. Migration decisions involve many aspects that are not quantitatively measurable, including perceived psychological costs and benefits. Therefore, I have adapted the cost and benefit model in my study to include migrant workers' perceptions of qualitative gains and losses.

Labour migration is no doubt economical. However, in analyzing migration predictors (in the previous chapters), it is obvious that migrants and their families construct an imagined world outside Bangladesh where jobs are available, life is secure and opportunities exist, although the situation in most of the receiving countries differs from such imagined constructions (Dannecker 2003). The factor analysis in my study shows that one of the basic reasons for migration is psychological. This means that economic factors alone do not necessarily trigger migration.

Neo-classical economic theories explain migration by referring to a push and pull framework as contributing to the conditions which make people migrate. Again, it is assumed that people migrate in order to respond to pull forces because they are purely rational economic actors. My study has revealed that, at the decision-making level, the push forces are overtly active, while the pulls are more covert. That means that migrants decide to move when they face the realities present in the country of origin, for example, unemployment, widespread poverty and family pressures, while the pull forces, i.e., the attracting factors abroad, are not known, except through their own expectations based on information or disinformation they receive via agents, brokers, friends, relatives, and the media. Therefore, the majority of my respondents claimed that they were ignorant about the realities abroad before they moved, and those who had received information from friends and relatives were better informed than those who had been deceived by agents and brokers, who, themselves, constructed pull factors using discourse.

I have highlighted the wage differential between the places of origin and destination. The study has clearly shown that migrants' income prior to migration is usually succeeded by their income after migration. This analysis supports neo-classical and macro-economic theories. The migrant workers' own realization of the actual wage differential is, however, *post facto* – i.e., it is often lower than their pre-migration expectation, even if it is higher than their pre-migration reality. As a result, they are disappointed by unmet expectations, even thought they realize that they may not be able to match even this lower-than-expected wage if they were to go home. They are thus compelled to stay on after they arrive, in a way that they were not before.

Relative deprivation creates desire for a better life standard. Therefore, migration becomes an option for people to fulfil their desires. Widespread economic disparity creates feelings of relative deprivation which have been influential in making migration decisions as a way to bring about economic parity. The theory of relative deprivation has not been applied in my research because the comparisons made by my respondents are not commonly between themselves and superior others known to them, but rather between their actual experiences of poverty and unemployment and their desires for waged work.

As individuals, migrant workers transfer remittances in order to develop the human capital of those left behind – e.g., through money sent for healthcare and education. This study has demonstrated that a significant portion of migrants' remittances is, in fact, spent on human capital development. Many migrant workers expressed being content that they could at least contribute to the human capital development of their family members. This again is more of a family capital strategy, rather than an individual's human capital strategy.

Generally, networks are understood by ties with friends and neighbours who have emigrated already or are helping their compatriots to immigrate, find jobs and adjust to a new environment. Based on this, a general assumption is made that these networks reduce the costs of migration for newcomers and further induce potential migrants to leave their countries. However, this study has revealed that networks are not limited to these types of personal ties and that the role that they play in facilitating migration is limited. I have extended networks analysis to discuss the role played by agencies, brokers and syndicates in migration facilitation. This extension has enabled me to grapple with the migration process of both destination countries i.e., Hong Kong and Malaysia. This study has confirmed that recruiting networks based on agencies, brokers and syndicates have, in no way, reduced the costs of migration and may even have increased both costs and risk in many cases. Thus, a basic component of networks theory has been challenged.

At the initial stage of migration, collective decision-making is seen as significant in the migration process. Therefore, household strategy theory is quite influential on migration decisions. However, household strategy theory has not been able to explain the fact that migration decisions can also be made individually. According to household strategy theory, a migration decision is made collectively by family members so that the risk can be spread. However, I argue that decisions are not always made collectively because in Bangladeshi society, the members of a family do not hold equal power. In addition, most men take financial responsibility and plan for household expenditures. Many women do not have any access to family decision-making. Therefore, collective decision-making is an immaterial issue in most cases. My study also confirms that many of the respondents decided to migrate against the interest of the family and, despite this, they still thought that their family members would share the risks they might face. Human capital theory asserts that people tend to migrate at a younger age as they have better labour market prospects than older ones. In my study, it was found that household heads may strategize to choose the most suitable family member to migrate for work overseas. This was therefore more of a family capital strategy by household heads, than an individual's human capital strategy.

After arrival in the host country, adaptation strategies are important for arranging a living place, obtaining work and settling work contracts, if any. Adapting to a new society for Bangladeshi migrant labourers who have grown up in a very different cultural milieu has always been challenging. Rationalization is embedded in the concept of adaptation, although it is not limited to cultural adaptation. Rather, it also entails economic, spatial, social, political, and global assimilation. I disagree with the proposition of assimilationist theory that adaptation is based on the intended length of residence in the host country. Rather, migrant workers plan their stay and adaptation in order to regain the money that they spent to finance their migration and to acquire a surplus. Adaptation is thus not the result of staying in the country, but a means of earning income. My study demonstrates that migrant workers adopt several adaptation strategies for this purpose.

While migration decisions are generally made with expectations and promises of better jobs and future prospects, with a few exceptions, migrant workers often fail to bring these hopes to fruition. Therefore, this study has dealt with the sources of decision-making and their subsequent mode of rationalizing their migration decisions by analyzing the many aspects related to their departure, routes, income-expenditure, remittances and experiences.

Although conventional understanding holds that migration is the preserve of demographers and geographers, and while many non-economic factors provoke migration, it is generally perceived as an economic phenomenon. Therefore, it is assumed that people migrate mainly for economic gain. Few studies contest the inference that migrants leave their place of origin due to a lack of employment opportunities and in search of fortune. The agricultural sector in most Asian countries has a limited capacity to absorb the growing labour force. Empirical studies have found that industrialization can create the capacity to absorb labour released from agriculture. Indeed, the number of job seekers in industrialized regions and countries has proliferated in recent years, mainly because of the stagnation of rural economies and the dwindling contribution of agriculture to the national economy. This has meant that the poor have been pushed out of the rural sector and pulled into industry and services. A powerful component of this absorption has been overseas employment. In Bangladesh, the relatively low level of industrialization has not generated adequate employment and in turn has contributed to a huge supply of potential migrants. Bangladesh's chequered political history, severe income inequality, high illiteracy, and rampant incidence of corruption has rendered formidable, if not impossible, the task of generating additional employment. Recent political unrest has also led to a sense of desperation among potential migrants from Bangladesh. Therefore, migration to overseas countries appears to be the best available choice for them.

However, while this may be the case, not all the poor migrate. Nor do many of those who migrate necessarily benefit as expected. In addition, ultimately, those who migrate may or may not return. Therefore, even among the poor, decision-making surrounding whether or not to migrate has to take place. Even after they leave, they have to make decisions about what kind of work to do, how to earn money, whether to send money home, how much to send, how to send it, as well as whether and after how much time, to go home. Therefore, poverty does not operate in an automatic, unmediated way to simply push people out. People have to decide whether to go and, after going, whether to stay away or come home.

This study has found that most of the respondents set off for overseas destinations with similar expectations and that they tended to rationalize their migration decisions in similar ways. The profile provided one common feature of the migrant job-seekers, regardless of their divergent preference for destination (either Hong Kong or Malaysia), in that they came from a similar socio-economic background, although those who went to Hong Kong seem to have more social capital than those who went to Malaysia. The similar socio-demographic background did not induce the respondents to end up at the same destination. In addition, their job status and income levels varied greatly. This study finds that geographical origins often characterized different migration patterns and, as a result, migrants from specific regions of the country with out-migration propensities might benefit from collective migration experiences as social capital, despite the identical domestic economy, political and social contexts between regions. Hence, the main source areas for Bangladeshi out-migrants are composed of only a few districts, including Dhaka, Chittagong, Comilla, Tangail, Sylhet and Noakhali. This finding is supported by the other studies (Siddiqui 2004). Migrants who are not from these areas lack the collective social capital and information to advocate for their migration, and are more likely to be cheated by agents and brokers.

The study has taken networks, routes and destinations into consideration because they have much to do with the migration process. This is why networks have long been recognized as significant to the migration process by many researchers (Faist 2000). The fact that the presence of relatives in both the places of origin and destination reportedly supports migrants in leaving the former and settling in the later indicates that they play a significant role in promoting migration decisions. This study has demonstrated how different types of networks shape the routes and destinations elected by migrants, as well as how they explain migration processes and experiences.

This study has broadly identified two types of networks (interpersonal and recruiting networks) that are operative in this particular migration process. The current study found that migrants use particular types of networks to get their migration trajectories facilitated. The network theory presupposes that networks reduce the cost and risk related to migration. This study has found that the majority of the migrant workers reached Malaysia through either Thailand or Singapore, relying heavily on recruiting networks for their migration. As a result most of them took weeks and some took even months to get to Malaysia. During this journey they suffered diverse adversities including starvation and threats to their lives. Therefore, this study has challenged the main proposition of the network theory: in these cases, the networks used reduced neither migration cost nor risk. In the case of the respondents in Malaysia (BRMs), the formal networks have instead increased the risk and cost of migration. The study has further revealed that the recruiting networks work at different levels, i.e., national, district and division levels. The study has found that the respondents in Hong Kong (BRHKs) did not rely heavily on such networks for their migration (in comparison to the BRMs), rather most of the BRHKs got to their destination (Hong Kong) directly by air from Dhaka and without the need to negotiate a visa beforehand. However, the migrant workers still considered networks in rationalizing their migration decision because migration cost is very much dependent on the types of network they relied on for their migration, e.g., interpersonal networks are likely to reduce the risk and costs of migration. The network theory is flawed in that it is limited in application to interpersonal networks.

According to rational choice theory, human beings opt for the best means by way which to attain their ends by calculating perceived future benefits, as most human behaviour is guided by reason. Therefore, economic gain through migration as a desired end could be achieved by taking a number of factors into account, such as costs incurred by and income earned through migration. Rationalization is done once migrants end the process and perform the cost and benefit analysis between the migration decision and post-migration impact.

The costs incurred by and the gains that could be recovered through migration were one of the primary concerns once the respondents moved. This study has made a comparative analysis of how much money was spent on and gained by migration. It was obvious that migrants spent a much larger amount of money than determined by the government. The study further revealed that those who migrate in illegal ways pay even higher amounts of money as the unscrupulous agencies tend to charge very high fees (Afsar 2003). Siddiqui (2000) adds that a quarter of the migrants were not aware of the regulations and maximum legally chargeable fees. The exorbitant migration costs they paid prior to their departure had an impact on their rationalization in the post-migration processes since many of the migrants took years to pay off the loans they had borrowed.

This study has demonstrated that the respondents in Malaysia spent significantly higher amounts of money than the BRHKs for their migration. The BRMs also borrowed higher amounts from various sources and took longer periods of time than the respondents in Hong Kong to pay back the money. There is evidence that the amount they pay to recruiters often becomes a debt payable in the receiving country. The terms of repayment are rarely specified but can include exorbitant rates of interest. Siddiqui (2003) points out that this debt often effectively binds them to their employers until the debt is paid off. Apart from the economic costs, the study identified other costs as well, including social, opportunity and psychological costs which cannot be easily measured using quantitative analysis.

The migrant workers were asked whether they regarded their migration decision as correct or not. An overwhelming majority of the BRMs reported that their migration decision had not been correct. Around half of the respondents in Hong Kong claimed that their decision to move to Hong Kong was correct. Most migrants rationalized their decision to migrate from an economic point of view which was linked to their desire to ensure financial security, not only for themselves but also their families left behind. However, despite their frustrations, there is no way to undo their decisions as they have already moved, spent the money, and are now indebted at home. Therefore, even if they regard their migration decisions as incorrect, they nevertheless decide to stay on overseas, as their concern is to regain the money they lost to financing their migration and they do not see any means of achieving this if they were to go back home. As a result, despite different evaluations of their migration decisions, as correct or incorrect, all my respondents rationalized the need to continue working in their destination countries.

Issues of living and working in host countries, income earned and remittance factors appear after arrival in destination countries. Therefore, these factors are included in rationalizing post-migration decisions.

Leaving behind friends, community, relatives and family places migrant workers in a challenging position. They remain in a very sensitive and fragile emotional state during the first few months after reaching their destinations until they get settled. At this stage, living conditions constitute a factor that they consider significant to justify their migration decision because this involves their level of income and how they are treated by their employers.

Adaptation comprises a number of aspects such as working and living conditions. In most cases, the accommodation is covered by their work contracts (in Malaysia). They often do not have any idea about the living and working conditions before they arrive, and many of them experience the actual conditions only upon arrival. Most of the BRMs were promised job packages that included free accommodation. This was an enormous incentive for them to make their migration decisions. This study has identified a number of different strategies that the migrant workers employed to adapt themselves to the different social and political settings in Hong Kong and Malaysia, including learning the local language and making local friends.

Migrant workers consider the working conditions in the host countries to be a significant factor in rationalizing their migration decisions. However, the working conditions of migrants vary, depending largely on the destinations and the type of employment. Generally, the majority of the migrants work as semi-skilled or unskilled labourers, with working conditions that are mostly substandard (Siddiqui 2004). However, the working environment at each destination differs. A key difference is how migrants obtain their jobs. For example, the BRHKs wait for daily work offers on the street at particular places in Hong Kong. They get job offers on a temporary basis and are paid by the hour. However, in Malaysia, the nature of obtaining work is different in that most respondents work on a monthly basis in shoe factories, paper mills, or on construction sites and at electronic industries. However, even though the workers in Malaysia have more regular jobs, the working conditions, as reported by the respondents, are more exploitative than in Hong Kong, as characterized by prolonged hours of work, lower wages, withheld payments, and sometimes brutal treatment. Therefore, although Bangladeshi workers in Hong Kong have irregular hourly work, they enjoy more autonomy and better working conditions.

The BRHKs lived in Hong Kong legally and worked illegally. While most of the respondents in Malaysia claimed to have valid visas, many of them confessed later that these visas are fake. With similar socio-economic and demographic mores and expectations, the respondents ended up in two entirely different destinations in terms of culture and economic parity. Therefore, their levels of achievement and satisfaction, fulfilment of objectives, and form of rationalizations vary widely, as it depends on how the individuals concerned perceive all these factors. These levels are largely dependent on the varied degrees of comfort, adaptability, and exploitation experienced. The BRHKs were found content with their migration decision to Hong Kong when the cost of living in Hong Kong was excluded from their consideration. By contrast, it is when the abusive and substandard living and working conditions are taken into consideration, that the BRMs regret their decision.

Income acts as the inductive force to keep up the spirit of the migrants to continue their stay abroad, despite the many adversities they might encounter. While the majority of the migrants (both BRHKs and BRMs) said that their income was much lower than they had expected or been promised, the earning was higher than they could get at home. Therefore, they rationalize their post-migration decision and continue to stay in the host country.

Income earning and remittances are intertwined because the volume of remittances depends largely on the income base and expenditure patterns of the migrant in the destination country. This study has added weight to the contemporary debate on remittances by highlighting two forms of remittance: cash and kind. The study found that remitting in kind has become more prominent recently. The current study data show that the migrants in Hong Kong sent more remittances in kind than did the respondents in Malaysia, as the BRHKs travelled to Bangladesh more often than the BRMs. Earlier, items carried by the migrants themselves were meant as gifts or for personal use. This study has discovered that nowadays these carried items are a means of transferring remittance. Easy transferability and tax free provisions on some selected items have promoted the intention to remit in kind.

One interesting finding is that as long as migrants stay abroad and remit money, the family in Bangladesh spends the money in a form of conspicuous consumption. In connection with this, many migrants reported that family members did not know how much pain they were enduring in order to earn money abroad. The lion's share of the money was spent on unproductive schemes at home, although a part of the money usually went to human capital development, such as healthcare and education. The respondents feel that this mishandling of money occurs due to their physical absence. It is, however, difficult to empirically confirm the extent of the impact that remittances have on consumption. In most cases, for both the BRHKs and the BRMs, remittances were used unproductively for consumption: the purchase of clothes, household goods, and so on.

This study has further contributed to the understanding of the channels of remittances that migrants use, including the fact that illegal migrants prefer informal channels to transfer their money, which involves greater risk. Formal channels such as banks were reported to be slow, ineffective and corrupt. This is important, as it is the government who can ensure better services in the banking sector, for example, by offering a competitive rate of foreign exchange in comparison to the curb market. The injection of remittances into the family economy is one of the best indicators of rationalized post-migration decisions to continue working overseas. Whatever way the money is spent at home, the migrants feel that at least their family members can live a better life. Many of the respondents said that they do not look back on the hardship they suffered; income for the purpose of remittance is their primary concern. Despite adversities at every level of the migration process, they still rationalize their migration decision as correct through the hope that they will be able to fly to some other, even better country and will be able to gain economic well-being there someday.

Migration and the public policy

Migration has become a key policy issue for Bangladesh, as in many other developed and developing economies (Skeldon 2003). This section therefore, critically discusses some policies held by the government of Bangladesh regarding the context of the current research. No pragmatic policies have, so far, been known to have been made to promote migration. For example, until recently the Bangladesh government's policy restricted women's migration, except in the case of professional migrants. With the overwhelming presence of foreigners in Hong Kong from the USA, UK, the Philippines, Thailand, Indonesia and Sri Lanka, the presence of Bangladeshis is minor.

Malaysia has the biggest market in South East Asia for Bangladeshi labour. A timely response to the demand for labourers in Malaysia was not made and, as a consequence, opportunities for Bangladeshi labourers in the Malaysian market are dwindling. Malaysia estimated a need for approximately 500,000 foreign workers in various sectors in 2005. The foreign workers cabinet committee (FWCC) made a list of 16 labour-sending countries but excluded Bangladesh. In addition, and despite a great deal of labour demand, around 150,000 Bangladeshis had to return to Bangladesh in recent years (Haider 2005, Hossain and Ullah 2004). Astonishingly, the labour sending agreement remained ineffective for more than seven years after it was signed with Malaysia. Finally, the agreement began to be enforced in August 2006. However, with the sedentary existence of the agreement, the governments' slow response and failure to fulfil some of the conditions, the implementation of agreement remains at risk.

While migration is gaining momentum due to its significant contribution in sustaining the national economy of Bangladesh, the gender dimension of migration from Bangladesh has received little attention. Bangladeshi migrants are predominantly men. This study further demonstrated that, despite the increased flows of women as independent migrants, the ratio between male and female is as wide as 100:0.98 (as of 2003) (Samren 2007). This finding can be attributed to the policy restrictions on female migration from Bangladesh. The socio-cultural and religious values of the country discourage women's free mobility, even within national borders, let alone migration to foreign countries.

Government policy was found to be deficient in providing even minimal support in the cause of promoting migration in Bangladesh (*Prothom Alo* 2005). Migrants' limited access to information, inefficiency, red-tape and corruption in the government machinery have discouraged migrants. This has resulted in the unabated growth of operations for organized criminal syndicates, particularly in trafficking women and children from Bangladesh. The flaws in government policy regarding migration have resulted in the exorbitant cost of migration incurred by migrants, the emergence of fraud by recruiting brokers, and the rising level of risk. The study also found that, in transferring remittances, the covert and corrupt practices in the national banking systems are redirecting the remitters toward the use of the informal *hundi* system. However, encouraging and ensuring cheaper, convenient and reliable ways of transferring remittance and the setting up of long term incentive programmes for attracting migrants as customers could be a viable policy intervention. Demand for Bangladeshi workers from overseas is processed mainly by the Government, private entrepreneurs and through relatives of individuals working abroad. There is a government-owned company, Bangladesh Overseas Employment and Services Limited (BOESL), and many private entrepreneurs have been given recruiting licenses, as per the procedure laid down in the Emigration Ordinance, 1982 (BMET 2007). The government established a Ministry of Expatriates' Welfare and Overseas Employment (MEWOE) in 2001 to ensure the welfare of Bangladeshi expatriate workers and explore overseas employment. The Ministry is currently rendering services in enhancing the flow of remittance.

According to the MEWOE, it has been arranging transportation facilities for expatriate Bangladeshi to allow for risk-free movement from Zia International Airport, Dhaka to nearer inter-district bus stations, to be used after arrival in Bangladesh. However, such arrangements were not available during the Author's four visits to the airport over the last two years. BMET arranges briefing sessions for the workers before departure to their destinations in order to make them aware of the existent agreement of service conditions, working environment, culture, salary and other benefits, remittance systems, and the local language of the host country.

This policy has included tax benefits that have been accorded to Bangladeshi expatriates. The National Board of Revenue (NBR) offered a number of tax facilities to expatriate Bangladeshis in 2002, some of which included a tax identification number (TIN) certificate, not required for buying immovable property in Bangladesh and the investment initiative for expatriate Bangladeshis in the field of business, commerce and industry. These existed for around three years and, until 2005, were accepted without question and without consideration to the amount of the investment; interest that accrued from migrants' non-resident foreign currency deposit was tax-free and investment in the agricultural processing industrial sector for expatriate Bangladeshis was also tax free until 2005 (BMET 2007). This means, however, that no long term incentives are currently offered.

These policies have meant that no functioning welfare system for migrants is in place in Bangladesh. Moreover, complaints abound that the migrants do not get required support from the Bangladesh embassies and consulates abroad. Common allegations against the Bangladesh Consulate in Hong Kong and the Bangladesh Embassy in Malaysia have included ill-treatment and misbehaviour by officials towards the migrant workers. In the wake of the growing prominence of international migration and migration policy development, very little has been done in Bangladesh to influence and shape the policy and modalities of migration for the benefit of migrants and of the country as a whole. Unlike neighbouring Asian countries, corresponding policy shifts regarding labour migration have not taken place in Bangladesh. On the contrary, the Bangladesh government appears to be insensitive and indifferent to problems encountered by migrants, both at home and abroad. Several recent incidents regarding the deaths of illegal Bangladeshi migrants abroad rocked the nation, along with reports about fraudulent recruiters' abortive attempts to send Bangladeshi labourers abroad with fake work permits. These incidents demonstrate the government's lax control over the whole infrastructure surrounding labour migration. Therefore, factors contributing to the rationalization of migration decisions are not merely an individual affair. Rather, a number of these factors are embedded in policy decisions.

The extent of the difference between expectations and forecasts that the respondents in Hong Kong and Malaysia held before migration, and what they were able to attain after migration, in absolute terms, remains very challenging to define. The study has demonstrated that respondents in both destinations have fallen far from their expectation thresholds. However, the gap appears to be wider for the BRMs. Therefore, dichotomous conclusions (both positive and negative) are postulated, as the gains are mostly tangible and the losses are intangible. When tangible gains are taken into account, migrant workers' opinion of their decision reflects a 'win-win' proposition. However, when the intangible losses are taken into consideration, their decision to migrate is generally deemed a loss of wealth, health and youth. With regards to the protection of migrant workers, the recruitment and preparation of migrants for deployment overseas, which currently remain the domain of the source country, could be more effectively handled by the government by preparing migrants thoroughly for what they might face in destination countries, in terms of both physical and mental stress, and skill-building to better self-advocate and deal with possibly predatory agencies and brokers.

Last but not the least, not all migrant workers are treated poorly; many are treated with respect and dignity. The culpability of employers in the abuse of workers can ostensibly be linked to the migrants' often illegal and degraded status. Therefore, it is necessary to focus on the vulnerability of migrant workers overseas and to seek mechanisms to reduce their vulnerability through policy decisions upheld by the governments at both ends, through bilateral agreements in order to manage migration effectively and combat illegal migration and trafficking.

Appendix A

Case studies: The return migrants in Bangladesh

Case study 1:
Mahbubul Murshed (27), assistant electrician in a Malaysian paper mill

Murshed came from Dhaka city, the capital of Bangladesh. He is the second of three children (the oldest male child) and both of his parents are still living. He completed a secondary level of education (12 years of schooling) and had six months training in basic electrical wiring from Dhaka Poly Technique Institute. He had been married for four years, without children, when he returned from Malaysia. He had been working as an electrical assistant in Muda Paper Mill SDN BHD at Kajang in Selangor, Malaysia for about six years.

Murshed's father, now retired from a private company had been the only income earner in the family. His retirement made Murshed consider the need to support his family. He had a sister who would need a dowry to marry and a younger brother who should continue his studies. With pervasive job scarcity in Bangladesh, Murshed knew that it was unlikely that a job would be available to him in order to support his family. Out of a sense of responsibility, as the eldest son, he felt that he had to find work. He tried to earn some money in Bangladesh by managing a video game shop, but earned only Tk. 500 (around USD 8) per month. As soon as he passed his college examinations he decided to leave the country in search of work.

Although Murshed felt apathetic towards his studies and wanted to work, his parents wanted him to continue with his studies, despite the economic hardship that the family was facing. However, one of his uncles was working in Malaysia at a recruiting agency which supplied manpower to various companies around the country. He informed Murshed that his company was going to recruit a number people from Bangladesh and proposed five aspirant migrants, including Murshed, to migrate to Malaysia. Murshed considered this a good opportunity and tried to convince his parents to allow him to go and to support his migration. Considering his reluctance to pursue his education his parents allowed him to go to Malaysia. In Murshed's case, his parents' consent was particularly important because he was still largely financially dependent on them.

When Murshed decided to migrate, he had been promised to earn RM 700 per month, which was a very attractive offer and therefore the cost-benefit analysis resulted in his decision to migrate. However, despite his uncle's involvement, he was deceived and his salary was reduced to RM 350 per month in practice;

although, with overtime work he was able to earn around RM 550. Murshed paid Tk. 65,000 to Shorker Recruiting Agency (SRA) in Bangladesh at Banani, Dhaka to finance his migration. The money was paid to them directly. His father financed his migration using his retirement benefits in hopes to earn back his investment many times over.

Murshed's proficiency in spoken English was poor, but nonetheless he was confident that he could communicate with people abroad. Indeed, he was able to pick up a conversational command of Malay in approximately three months. Prior to leaving for Maylasia, Murshed knew that it was a Muslim country and that there were a lot of Bangladeshi people living there, so he reported feeling unafraid and comfortable. The company that hired him provided him with accommodation (in a worker's dormitory) two kilometers away from their factory and a bicycle to commute to work every day. Most of the other inhabitants of the dormitory were Bangladeshi and they would cook indigenous meals for themselves since they were able to easily find the ingredients at the Sungai Rama temporary market.

Murshed spent approximately one third of his salary on his own subsistence and the rest he sent back to his father through *hundi* at intervals of about three months. His father reportedly used the earnings to buy government savings bonds from the General Post Office. Murshed's father used the interest received from the savings bonds to maintain the family.

Since Murshed lived in the capital city of Bangladesh he was exposed to newspapers and electronic media, and was exposed to information about Malaysia through these sources. He knew that Malaysia was a Southeast Asian country and not less developed than Bangladesh. He said that this information also played into his decision to migrate to Malaysia. However, for the more practical information that he required, he relied on his uncle. Murshed reported that he had to decide carefully on whether and where to migrate because if he was deceived there would be no way to recoup the loss. The entire family had counted on and trusted the information received through his uncle and it was the direct involvement of his uncle in his migration process that had made him confident that he was not going to be deceived.

Despite his deception, Murshed did not regret his migration. While in Malaysia he was motivated by the desire to bring back the well-being of his family. However, he reported that homesickness sometimes weakened him. When he tries to balance the equation of the money earned and the six years of his life that he spent labouring there, he says with a long sigh ...

> ... something is better than nothing. Yes, six years is a pretty long time that I spent abroad. But what else could I do here in Bangladesh? To me, it was worth staying in Malaysia, at least earned some money, which I cannot think of in Bangladesh. I don't regret my migration. If I have a chance I shall remigrate.

To Murshed, it seems that the economic, political and social conditions in Bangladesh are getting worse than ever. When asked what he means by these conditions, he anxiously replied that the salary he earns here is one-sixth of that he earned in Malaysia, political stalemate causes the loss of work days and a lack of dependable social security. He believes he could help his family better if he could remigrate. He is ready to invest the capital [currently in the form of the savings bond] for remigration.

When asked whether he faced any ordeals during his stay in Malaysia, Murshed said that he had had to turn over his travel documents to his company and a photocopy of the documents was given back to him (a common practice). However, he reported that police sometimes wanted to see the original copy even though they knew that his passport was probably with his company, On more than one occasion he was held for long time on the suspicion that he was carrying false documents. He was only able to secure his release at the cost of bribe; sometimes RM 5, sometimes more. With a heavy tone of voice he stated that all of the police were of Indian origins. He still wonders that Malaysia could be an example of communal harmony. Despite this, Murshed says that many ethnic minorities stay in Malaysia with complete congeniality and that he likes the country very much.

Case study 2:
Anowar Hossain (32), a worker in a Malaysian Paper Mills

Hossain, single, from Barisal district some 160 km south of Dhaka city, was second among his two brothers and three sisters. He had 10 years of schooling with no professional training. He wanted to continue his studies but could not because of financial constraints. Without any income sources, Hossain's father had to depend financially on his eldest son who was reluctant to let Hossain continue his studies.

> ... it was the hardest time of my life. Idle time is the hardest time when somebody is living in poverty. My brother and his wife indirectly pointed out many times that I have been a burden on the family and that they would no longer be able to maintain my expenses. Misbehaviour both from my brother and his wife exceeded the limit of tolerance. My father had no say in the family because he had no income. I understand in my heart now that money is everything. It was better to die than be a burden on others. I decided many times to commit suicide, but I could not because it is forbidden in Islam. About four years I had to pass my time in the same manner ... One of my uncles responded to my request and gave me information on a job offer abroad. I told him I would accept whatever and wherever the job is, even if it is in the "hell"! Therefore, I accepted the offer.

Hossein's uncle helped him to process his migration, and gave him a great deal of information regarding the cultural mores of Malaysia, which was later augmented by his friends and other relatives. Hossein's decision to migrate was forced by the adverse economic conditions his family faced, and his brother's unsupportive attitude. Nevertheless, his brother discouraged Hossein from migrating to Malaysia; Hossain claims that his brother did this out of jealousy, because his brother thought that Hossain was going to be rich.

Hossain spent Tk. 75,000 to finance his migration. He had to borrow the money from his relatives, and his brother did not extend any support. He gave the money to a recruiting agency directly. He was lucky and received an offer from Malaysia, which he knew was a Muslim country. He was glad as this was better than he had expected, since he had been ready to go anywhere in the world.

The main motivational forces behind Hossein's migration decision were his desire to support his elderly father and change his unemployment status. Hossain worked in Malaysia for five years as a labourer in Muda Paper Mill SDN BHD, at Kajang in Selangor. His salary was RM 450 per month, which was what he had been promised. Almost half of the money was spent on his subsistence and the rest he sent home every four months, through other workers travelling back to Bangladesh or through bank transfers.

His father saved some money from the portion that was sent to him. With it he bought some property. Since returning, Hossein has again become unemployed. He is planning to remigrate to any country in which he has an opportunity. However, he is concerned about the increasing trend of deception taking place in the recruiting processes. He mentioned some recent incidents of deception on the part of the recruiting agencies and the fact that, as a result, many aspirant migrants had been deported from their destination countries.

Case study 3:
Mokaddes Hossain (30), a labourer in an Electric factory in Malaysia

> ... it was the most disgraceful work I ever did in my life. I never thought of doing such types of work. People with no other choice should do things like that. I was simply forced to go abroad by my guardians. However, while abroad they thought I was going astray. My employer never considered us as human beings.

This was how Hossain, the fourth of nine siblings, expressed his depression and anger regarding the five years that he spent as a migrant labourer in Malaysia. In his view, this period of time had destroyed his confidence and zeal for work. He felt that his guardians had wanted to become rich through his income. Hossain had had 14 years of schooling and reported being a hardworking and innovative man. He had been confident about finding a suitable job for himself in Bangladesh, claiming that he had a level of education that would allow him to be a manager there.

> ... I was volunteering in a factory owned by one of my friends with a view to having training on export-import and garments trade and I used to go back home late and the family members thought I was doing otherwise.

However, according to Hossain, he had no say about his future and was sent abroad against his will by his guardians. Once in Malaysia, he did not feel uncomfortable in terms of the weather or the culture. Moreover, he felt that Malaysia was not too far from home, as it was only three hours travel distance. It was the working conditions that had been so frustrating to him.

Hossain knew a reasonable amount about Malaysia prior to his migration, including historical and economic details about the rapid transformation of the country into one of the most productive economies in Southeast Asia. This was, in part, why he did not strongly oppose his family members' decision to send him there for work, a decision which he now regrets. Hossain feels that the five years he spent were futile as the work experience he gained is not applicable in Bangladesh and instead is viewed as an interruption in the continuity of his work at home. He has faced problems readjusting to the Bangladeshi working environment, which he now finds unpredictable.

Hossain spent approximately Tk. 70,000 to finance his migration. His migration was processed by a government registered manpower recruiting agency with a well-regarded office in Dhaka. He understood that the dangers of using an unregistered or temporary agency; and made sure that he used an established one. He paid once he had confirmed that he would not be deceived. Obtaining the required money to finance his migration had been the responsibility of his guardians.

Hossain paid back the money borrowed from his guardians within a year after he migrated. He did not send any additional money to them while in Malaysia. His salary was RM 500, which was what he had been promised. He generally spent approximately half of his salary for his own subsistence and the rest he saved with a Malaysian bank. The company he was working for provided accommodation for him in a workers dormitory with a group of fellow Bangladeshi workers.

Hossain expressed his regrets and disappointment with his overseas job because he cannot find a job now in Bangladesh, particularly as he had not been able to complete the training he was engaged in before his move. To him, things have changed a lot as a result of his migration. The work experience he gained as a labourer in Malaysia is of no value and he has now become unemployed. He invested the money he saved from his earnings in a savings bond, but the interest he receives does not cover his subsistence needs. However considering his current unemployment status and the scarcity of the job market in Bangladesh, he would consider remigration. He says:

> *Ehon edesh aar amgo chaina, aar paribar o na* [Neither this country nor the family need me any longer].

Case study 4:
Abdul Kuddus (37), a labourer in Hong Kong

Kuddus came from the district of Comilla. He was the eldest among three siblings and was married at the time of his migration. There was no other income source in his family: His father had become too old to work and his younger brother was still in school. Kuddus' only son was growing up. He felt that expenses of the family were too diverse to be supported by its resources, particularly since the family did not own much land. He had attempted to go into petty business before but had incurred a large loss and thus was reluctant to try this avenue again. Due to these circumstances, he began seeking overseas employment, without much consideration at to where or what the job was. He contacted one of his relatives staying in Hong Kong and asked him to contact a broker in Dhaka to facilitate his journey to Hong Kong.

> ... I was undecided on the point of deciding to go to Hong Kong when I heard from the broker that there was no need for a visa endorsement, no work contract and no visa for a longer period. But he kept pushing and alluring me. I called my relative in Hong Kong. He gave me the green signal! And I took the risk. No risk no gain.

Thus Kuddus set off for Hong Kong, intending to enter as a tourist and then look for work. He paid Tk. 70,000 to the broker, who briefed him on some possible questions the immigration officers might ask at both ends. He had to pay an additional Tk. 500 bribe to the immigration officers in Dhaka. He reported that at the Hong Kong airport the immigration officers were satisfied with his answers and he was allowed enter Hong Kong for three months as tourist. He was received by his relative at the Hong Kong airport and taken to a living place on Yu Chau Street in Sham Shui Po where four other Bangladeshis were living together in a single room. There he slept in a four storey bed (cage home). His relative taught him some strategies for obtaining work, including possible streets which were best to wait on in order to get a work offer, how to recognize people coming with a work offer, how to do the work they needed, how to satisfy Hong Kong employers, and how to answer the police if he got stopped. His relative also instructed him on how to negotiate wage rates with the employer and to understand the entire work situation. He was told that when a car slowed down on the turn, he should rush to the car and ask for work. Later he learned more about the legal issues of work, including how to obtain a Chinese visa and get the visa renewed from Shenzhen, and how to use multiple passports. He reported learning Cantonese very quickly and eventually getting more work than his relative did.

After a few months he thanked his relative for encouraging him to come to Hong Kong. He said there was risk everywhere but that the money he earned was commensurate with the risk. He explained an ordeal he faced in which, one day, he forgot to bring his passport with him and while waiting for a work offer at Sham Shui Po a police officer asked him to show his travel document. He was detained but later one of his friends brought the passport and got him released. However, despite the possible difficulties, he was earning on average HK$10,000 per month which was equivalent to what he spent on his migration initially.

... I got money but I lost my father. News came to me about my father's death. But I could not go back to Bangladesh to see the face of my father for the last time because I knew if I went I could not come back again. With a tourist visa for three months I stayed more than three years, on visa extensions. The immigration officers are not foolish. I made my heart firmer. As I have lost my father already, I have to stay some years and earn a sufficient amount so that I can do something worthy in Bangladesh when I go back for the future of my son.

Kuddus would spend an average of HK$2,000–3,000 per month and the rest he sent back to Bangladesh. Part of this amount was spent on family maintenance and the rest was saved in the bank. He used to send the money through a luggage party, which he found to be quick and efficient. With his earnings, Kuddus renovated his house and bought some land. However, his mother eventually fell sick and this prompted him to decide to return after eight years of work. After the handover in 1997, work opportunities in Hong Kong reduced and competition become keener, which resulted in declining payment. In addition, tourist visas were reduced to two weeks and additional legal hassles discouraged him from continuing to prolong his stay in Hong Kong.

When Kuddus returned to Bangladesh he had saved a few hundred thousand Taka. He used this to start a business in his district town. Currently his son is going to school and his mother is well taken care of. He said he lost eight years of his life working illegally, but gained a lot as well.

Case study 5:
Yousuf Ali Patwary (37), a labourer in a textile mill in Malaysia

Patwary, the eldest son among four siblings, came from Gazipur district, 10 kilometres from Dhaka. He had 10 years of schooling with no other training or experience when he migrated, and was married with one son. His decision to migrate was based on purely economic reasons. He was afraid to migrate to a country very far from Bangladesh because he was worried he would not be able to afford to come home if he needed to. Therefore he had preferred Malaysia as a destination country from the beginning. He as also felt that Malaysia would be the best choice because it is a Muslim country and the weather would be similar. He decided on his own to migrate, and then got his father's consent.

Patwary paid for his migration in two instalments. A broker, who simply introduced an agent, cheated him for Tk. 15,000. He then had to pay Tk. 75,000 for the agency to arrange his migration. He collected this money by selling off his land assets.

... I went mad when I sold my land property and was deceived by the broker. I can never gain back the land. I become desperate to go abroad- anywhere. I thought if I couldn't go anywhere I couldn't buy back my land, which was our inherited property. No one wants to lose his ancestral assets. Nevertheless, I did.

Patwary worked in Malaysia for about five years. His contract had been for RM 700, and he was surprised when, upon arrival, his salary was significantly less (RM 450). However he could not do anything except accept the new terms, reporting that the other workers at the company had had the same experience. He worked more than 12 hours a day and used to work on holidays for extra income on top of his monthly salary.

The company gave him accommodations three kilometres from the factory and he would be transported to work by company shuttle. He ate in a mess hall with other Bangladeshi and Indian workers. He used almost half of the money from his salary for his subsistence and the rest he would send to Bangladesh through *hundi* every three months. He sent money through bank transfers only two times because the experience was not good (it took long time to arrive and the transaction cost was very high).

Patwary said that the money he kept as savings was not sufficient to do something good enough to support his family. He bought some land and was running a petty business in his locality. He did not feel that it was worth it for Bangladeshi people to migrate. He said that he could earn the same amount of money in Bangladesh if he were given the same job and worked as hard as he did in Malaysia. On top of this, he could stay with his family, which would be less expensive. He ruled out the possibility of remigration, even if a chance comes his way again.

Case study 6:
Syed Faizul Islam (39), a labourer in a paper Mill in Malaysia

During the interview, Islam showed a picture of a piece of land he sold in order to finance his migration to Malaysia. He had sold everything he owned for the dream of a better life. In the end, when he came back from Malaysia, he could not buy the land back. Islam, came from Noakhali district, and had worked in Malaysia for more than five years at the Muda Paper Mill in Selangor. In Islam's case, his sisters and brothers played an important role in his decision to migrate. He has four brothers and three sisters, and is married with one child. When he decided to migrate his brothers and sisters were studying, and part of his decision to go for an overseas job was to let others in the family finish their studies. He found information on the job in Malaysia from the newspapers and then contacted an agent. He paid Tk. 120,000 to the agent who kept him waiting for several months.

Prior to his migration, Islam had had a small business in Bangladesh. However it could not support the entire family, leading to his decision to migrate. His elderly father did not give psychological support because he was afraid that if Islam went abroad he would not see him again in his lifetime. Instead, he urged Islam to expand his business and stay home. However, despite this and a number of other hindrances Islam left for Malaysia.

... my dream was broken when I found my salary was less than half of what had been promised. I regretted going there and disregarding my father's urge. But there was no way to go back.

He was promised approximately RM 1,000 per month, however his actual salary was only RM 450. The amount he saved was even less than he used to earn from his petty business at home. The agent promised that he would be able to earn a lot through overtime work, which was not accurate. He had been happy to secure the job in Malaysia, but in the end it gave him only frustration. Almost half of his income was spent on his own subsistence. Every four months he would send money back to Bangladesh either through a bank transfer or using *hundi*. Currently he is unemployed. The capital he saved is not enough to re-start his business. Additionally, his neighbours thought that he had brought back a lot of money with him from Malaysia and rural *mastans* (hooligans) began hassling him, demanded money.

Case study 7:
Iqbal Ahmed (36), worker in a Coca Cola bottling company in Malaysia

... even as a BSc holder I had to work as a laborer with illiterate people. That was disgraceful for me. I was not treated as an educated person. There was variation neither in salary base nor in treatment. When my father could not bear the burden of our big family, I had to take over the responsibility and hence decided to migrate.

Ahmed worked for about seven years in a Coca Cola Bottling company in Malaysia. He has four brothers and four sisters and is a BSc graduate with computer diploma. He is still unmarried. He had been trying to find a job in Bangladesh and had been offered one, but the prospective employer had asked for a large bribe and Ahmed had thought it was better to migrate than to buy a job for a bribe.

... paying a bribe for a job to me is an insult to my education and to my own dignity. I did not want my career to be dishonestly started.

In addition, Ahmed needed for his migration less than half of the amount demanded as bribe for the job. He spent Tk. 100,000 to finance his migration and, although he was cheated by the recruiting agent, to him it was a better decision to look for work in Malaysia.

He received information about manpower recruitment in Malaysia from the newspapers and directly communicated with an agent, submitting his CV to them. He was sure that the agent was registered with the government. He received orientation training on the rules and laws of the country before he migrated and he had good conversational skills in English. He financed his migration using his own savings, substantiated by money from his parents. He did not need to borrow money from outside his family. His relatives encouraged him to look for an overseas job. He had an additional offer from behalf a bride who wanted to finance his migration as part of her dowry money, but he refused.

In Malaysia Ahmed earned approximately Tk. 28,000 per month, which continued to increase over the period that he was there and he was happy with this salary. He spent almost a quarter of his salary on his own subsistence and the rest was sent back to Bangladesh through *hundi* every two months. He felt that *hundi* were quicker, less expensive options which required less paperwork than bank transfers.

Ahmed's salary had to maintain his whole family, hence he could not save much to put towards future investment. Whatever he did manage to save is being used for his current small business. He does not regret his migration and reports that his company was very good to him and that his salary was greater than he had expected. While in Malaysia, he stayed in an apartment provided by the company. He cooked for himself in order to have indigenous food and this also reduced his cost of living.

Case study 8:
Ahsan Ullah Khandker (35), a labourer in Hong Kong

> ... days now are not same as before. I have heard from my friends that they are having in multiple problems in Hong Kong. What I earned in seven years they cannot do in seven decades! I don't feel like migrating again. Now I am fine here.

Khandker migrated to Hong Kong in 1995. He had two brothers and one sister, and was married with one child. He stayed in Hong Kong for approximately seven years on tourist visa which he kept extending by exiting to China and often Bangladesh, and then re-entering Hong Kong. He said that during this period normally three months of extension was approved. He chose Hong Kong as his country of destination because of the high level of income reported. He went there for the first time on a trial basis, thinking that, if he found convenient, he would extend his stay and if not then he would go back. In the end he remained there for seven years.

The information Khandker received about Hong Kong from some of his friends and relatives helped him much to make his decision. In 1995, he spent around Tk. 60,000 to finance his migration. This covered his air ticket and the charge for the brokers who helped arrange the air ticket and other documents. He dreamed of being a rich man as he felt that the society respects rich people, and not the poor. He had tried running businesses of many kinds in Bangladesh, but he could never make a profit. He was so desperate to move out of Bangladesh and he was willing go wherever he got a chance. He did not care about the consequences, saying 'whatever would be, would be'.

He asked for the money to finance his migration from one of his relatives who turned him down because they thought that he could not pay back the money. However his friends extended help and lent him money '… by the grace of the almighty I paid the money back to my friends within six months' he reports. He says he is still grateful to his friends. His father also helped him to finance his migration.

Khandker reports that he was never unemployed in Hong Kong. His average income was around HK$15,000 per month. He used to send his money to Bangladesh mostly through a 'luggage party'. He said that luggage parties were more reliable and speedy. Some times his family needed money immediately and therefore he could not use the bank transfer system because it was very slow. His sister and brothers were studying and his parents were sick and had to have medicine. Money he remitted was mainly used for educational purposes and for healthcare.

He bought land after he returned to Bangladesh and is now running a big shop in Chittagong City. He said he invested his money in his shop. He said that he is now earning a good amount and he is well-off. An excerpt from the interview follows:

> … I made a good decision which I never regret. I am grateful to the people of Hong Kong. It was a risk to go aboard without the guarantee of a job; nevertheless I knew without a risk I could not expect any gain.

Case study 9:
Joynul Abedin (33), a worker in a cargo company in Malaysia

Abedin worked in Malaysia for more than 10 years, for a number of different companies. Feeling that he was a burden on the family, his migration was primarily aimed at bringing financial relief to the family. One of his friends was acquaintances with a broker who promised to export Abedin to other countries, who had a verbal contract with him. The contract was made for three hundred thousand taka (Tk. 300,000) for migration to any developed Asian country, except Japan. There were a few other choices, including Tk. 500,000 for any country in Europe and Tk. 600,000 for migration to the USA. Abedin chose the cheapest option. The contract was flexible in the sense that he had to pay 70 percent of the total prior to his migration, and the rest could be paid once he reached the destination country. The broker travelled with him until he arrived in Thailand and obtained a visa for Singapore. Abedin said that the broker was very clever and smart. He had visited many countries and his passport was thick with visas from many different countries, so that the embassies or consulates did not ask him much upon visa approval. However, Abedin had expected to work in Singapore which was where he had been sent, but due to strong police surveillance, he was detained by police for not having a valid work permit. Afterward he decided to move to Malaysia.

Abedin's route was different from those who migrate to Malaysia legally. He migrated to Thailand first and was kept in Bangkok by the brokers for two weeks in a hotel close to Pratunam. He obtained a Singapore entry visa from Thailand for only one month. Once he moved to Singapore he understood quickly that it was not a good place for him to work and earn enough to compensate for what he had spent to finance his migration. He expected some help from other Bangladeshi migrants that he knew in Hong Kong, but did not receive it. He searched for other options and finally obtained a Malaysian tourist visa from Singapore and moved to Johor Bahru in 1995. This was a positive turning point for him: He obtained work in a Chinese company and the owner of the company protected him from police detention because the owner was able to offer him a lower wage than local labourers. Abedin was happy with the arrangement.

Abedin's was a particularly expensive migration. When asked how he was able to raise the amount of money necessary he indicated the land around the pond where he now raises fish and said:

> ... that time, I sold that portion of my parents' land but look (indicating the other side of the pond) I have bought back more. The land you see in front of you is mine. Look, look at my house, it is one of the best houses in our locality. My younger brother studies in Dhaka, and I finance his study. I look at the graveyard of my parents, it is cemented. I did this. All you see is from Malaysian Ringgit. It is all from my Chinese employer. I am grateful to him, he sheltered me. However I was sincere as much as possible and I was truly rewarded.

The sale of his parents' land alone could not entirely fund his migration and in addition he had had to borrow money from his relatives. He took two years to repay the loans but he was able to do so, and he bought (he does not want to say bought-back) his parents' land in 2003 upon his final return. He currently is not thinking of remigration, saying that human life is short and he needs to spend it with his family. He feels he has responsibilities at home.

Abedin remembers some of the ordeals he faced during police raids in Malaysia. He had to hide himself inside his company's building, facilitated by the owner. He said there were a number of Indonesian and Nepali co-workers who were also illegal and that the owner sheltered all of them. He feels that both parties benefited from this evasion strategy. It was his ultimate goal to be financially established and he feels that his migration allowed him to achieve this.

An excerpt from the interview illustrates how desperate he was to migrate:

> ... during my departure from my country, my relatives and friends bade me farewell at the airport and I said to those present that this Abedin (pointing his thumb at his chest) will be back with money or without life. I returned with the former.

Case study 10:
Mokadder Hossain Mithu (30), a worker at ASE Electronics Company in Malaysia

Mithu spent approximately six years in Malaysia, working for ASE Electronics Company. He was one of six brothers and four sisters, and had had ten years of schooling without any additional training. One of his brothers, who was already working in Malaysia, persuaded him to migrate there. He had failed his Higher Secondary Certificate examination in Bangladesh and his deteriorating relationship with his eldest brother forced his move. He knew about Malaysia and manpower recruitment there from his brother. Since his brother was involved in the recruitment and migration process he was confident that he would not be cheated.

Mithu asserts that he was sent by his family members. He failed his HSc examination twice and felt apathetic towards continuing his education; therefore they felt that migration was his best possible choice. His brother had married a Malay woman as a strategy for adaptation to the society and intended to stay in Malaysia permanently. However, the Malaysian law does not allow a migrant to stay permanently despite his martial status.

Mithu spent Tk. 80,000 to finance his migration, which he borrowed from his brother and repaid within one year while he was in Malaysia. He was paid approximately Tk. 20,000 which was what he had been promised. He generally spent one-quarter of his earnings on his subsistence in Malaysia. The rest he would hand-carry to Bangladesh when he visited once a year. He seldom used formal channels (e.g., banks) to transfer money.

The savings Mithu earned were not enough even to start a petty business in Bangladesh. Therefore, he is trying to invest his capital for remigration to another developed country. He reported that he had already given his savings to a broker who promised to take him to Canada. He said that he gave him the money in early 2004 and the broker assured him that within four months they could depart. At the time of the interview Mithu was worried as almost a year had passed with no positive response from the broker.

> ... but now I have training as a chef from the Porjoton Corporation [a tourism
> corporation]. I had this training because I am always thinking of remigration. So
> far as I know, there is demand for this in Europe. There is no future in Bangladesh
> especially for those not highly educated. Migration is my best choice.

Mithu's opinion on Malaysia and the behaviour of the Malaysian people was different from many of the other migrants interviewed: he felt that citizens there abuse Bangladeshi labourers by using pejorative names such as *Banglalees beggars* etc., in their own language.

Case study 11:
SM Quaruzzaman (34), a worker at a construction firm in Malaysia

Quaruzzaman spent one year in Malaysia. He had had ten years of schooling and had been working as a labourer with a construction company in Bangladesh. He was also trained in motor mechanics.

Quaruzzaman was the third of seven siblings, four brothers and three sisters. He is still single. After he failed to arrange a job in Bangladesh he became frustrated. His education did not help get a job and his family was facing financial hardship. Under these circumstances, one of his friends advised him to talk to a broker he knew. The broker was famous for helping people to migrate from Bangladesh successfully and he was known to be trustworthy. Quaruzzaman thought that this was not a bad idea, and decided to meet with him. The broker told him that he could buy a visa for Malaysia and that it would cost Tk. 120,000.

Quaruzzaman was suspicious because the amount was double what he had heard was normally required to buy a Malaysian visa. He told the broker about some examples of others in his village who had migrated at the cost of Tk. 60–80,000, however the broker insisted that the price was now Tk. 120,000. Quamruzzaman was not sure where he could collect that amount:

> ... I was thinking just like a madman how to arrange the money. I convinced my other brothers and sisters to sell the piece of land we all share. My brothers and sisters consented with the hope of seeing their economic condition improve. But I had to sell the land at half its price.

As soon as Quaruzzaman handed over the money, the broker went into hiding. After six months he appeared in the village and with the help of a local arbitrator Quaruzzaman got Tk. 100,000 back with a loss of Tk. 20,000. Later he came to know that his friend got a share from the broker. He was shocked at this betrayal and this incident made him obstinate that he must go abroad in order to teach them a lesson and not be defeated in the game. He came to Dhaka and found an established recruiting agency which managed his migration to Malaysia after three months at the cost of Tk. 80,000.

> ... but salary was only RM 400. I got a second shock. I was promised RM 650 per month. I understood that I had made another mistake. I forced myself to stay there for one year and I did but returned almost empty-handed. I failed to keep my promise to my brothers and sisters.

Quamruzzaman said that he had lost one year but that he must gain back everything. He is now determined to get back the land he sold. He is currently running a shop that he started with money he saved from his work in Malaysia, as well as some borrowed from his relatives. He said that so far it had been excellent. He had learnt from his mistakes:

… we can earn more than we do abroad with the same amount of labour and time devoted, whether we have an education or not, as long as we are laborious and sincere.

He is no longer frustrated: His brothers and sisters are studying and he wants to see them educated. He wants to transform the family into an educated one. Here he finds satisfaction. He urges potential aspiring migrants not to believe any brokers because most of them are fraudulent.

Case study 12:
Mohammad Rezaul Karim (40), a worker in Hong Kong

> … going out of the country was the only option left for me. My political opponent came to power and declared openly to take revenge on me. I knew that the administration would no longer be with us. We tried to build up a consensus to maintain a harmonious environment together but failed: I am sure they have not forgotten the repression we perpetrated on them when our party was in power. Therefore, when threats to my life continued to come, I could not merely ignore them.

Karim was receiving information that a lot of violence was taking place in the community and that many of the victims were his peers. He did not feel that his locality was safe any more. Karim was clear that he left the country to take shelter. He had been involved in national politics in his locality and had become the general secretary of the party in his area. He used to arrange work by the Department of Roads and Highways to repair local roads and culverts etc., and since his party was in power they used to get all the work, under the auspices of his local MP, while his opponents had totally been deprived of work. He earned a lot of money during his time in office.

When he heard about the violence, Karim consulted some of his friends, local leaders from his party, on what to do, given the situation. They told him that they were going to hide, hence he decided to do the same. He thought it would be easy to go to Hong Kong and immediately decided to try. As soon as he got to Hong Kong he understood that he had to look for work, since it is one of the most expensive countries in the world, and as such his money was exhausted sooner than he had expected. He changed hotels every week to find the cheapest ones. He tried to make friends with other Bangladeshi labourers. He explained his position and sought help from them. They told him that he could do what they did i.e., work as a daily contract labourer or on an hourly basis. He agreed, as there was no other way to survive.

In the first instance, he felt insulted and did not want to accept the job. An excerpt from the interview explains why he felt offended:

> ... I used to employ workers while I lived in Bangladesh however the situation here was totally the opposite and hence my hesitation. Mentally I could not accept the work offers, but there were no other options left for me.

Karim did not think of saving or remitting money. He worked only for his own subsistence. He left enough money in Bangladesh to maintain his family, including his six siblings. He used to earn on average HK$6,000 and used it all on living expenses. He said that he saved more than he expected, but then began to smoke and drink a lot and in six months he managed to get an Indonesian girlfriend. He said that it required a lot of money to maintain his lifestyle. Although he was in touch with his family frequently over mobile phone he felt insecure about his kids and other members of his family because they continued to receive threats, but Karim had been the main target. He spent around Tk. 500,000 in four months. He had problems with the food because he did not like the food served in Hong Kong restaurants. Later he came to know about some Bangladeshi and Indian restaurants in Tsim Sha Tsui. He rationalizes his migration by saying that it saved his life. He says he also learned about Hong Kong and about different cultures, and had some good friends afterwards. Overall, it was a nice experience for him.

Case study 13:
Jahorul Hoque (38), a worker at the Sentiment Latex Company in Malaysia

Hoque, one of four children and the only son, had 10 years of schooling, and worked as a labourer in Malaysia for approximately six years at the Sentiment Latex Company. His father had become very old and was unable to work and no one else was able to look after the family financially.

When Hoque's sisters reached an age where they should be married (a ceremony which requires a lot of money even if no dowry is demanded) his father asked him to do something for the 'fragile' family. Hoque was at a loss as to what he could do without a good education and no savings. He spent a year as a vagabond. Friends and other young people teased him. He decided to look for overseas work, but at the same time was worried about the money required to finance the visa and how to gather it. He was well aware of the fraudulent practices of many of the manpower recruiting agencies and felt that he could not trust anyone to help him.

He talked about his decision to his would-be wife who suggested that he to talk to his her father. Hoque said that, although he was hesitant to talk to him it turned out to be a really good suggestion. He knew he must find a better option for his future. He spoke with his would-be father-in-law, who money him the money on the condition that he had to legally wed the man's daughter before he left and perform the social formalities after he returned to Bangladesh. Hoque replied that he would have accepted any condition other than death.

He thanked his would-be father-in-law and told his family that he was leaving Bangladesh, explaining how money was being obtained. His family was happy. His would-be wife wept at the news and Hoque felt that it was his inspiration:

after all, it had been she who made his dream reality. To him the six years he spent in Malaysia were like six decades: Time passed very slowly. He came back to Bangladesh in the late 2004 and finally brought his wife to his house. He has given half of the money back to his father-in-law because he knew if he did not, it would become a dowry. Hoque used the rest of the money to start a small business in his locality.

Hoque spent Tk. 100,000 to finance his migration. He said that this was OK because, fortunately, he had not been deceived at either end. His employer was nice to him, his salary was good enough, and the best thing was that no false promises were given. He was paid RM 600 per month, which he said was fine for him. Hoque saved money every month and sent a portion back to Bangladesh to maintain the family he left behind. He was happy. He sent money back though *hundi* as he fount it to be speedier and cheaper than the other available channels. He once tried a using a bank transfer, and with some anger reported that it was a mistake he would never repeat.

Appendix B

Some policy issues of migration

Hong Kong

In the area of emigration, the Hong Kong government follows a laissez-faire policy of non-intervention. Hong Kong residents are free to move in and out of the territory, and they do not have to declare the purpose of their movements to the authorities. This freedom of movement, much valued by the local population, has been enshrined in the Sino-British Joint Declaration about the future of Hong Kong beyond 1997. In the Basic Law, a mini-constitution for Hong Kong after its reversion to China, it is stipulated that:

> Hong Kong residents shall have freedom of movement within the Hong Kong Special Administrative Region and freedom of emigration to other countries and regions. They shall have freedom to travel and to enter or leave the Region. Unless restrained by law, holders of valid travel documents shall be free to leave the Region without special authorization.

However, a distinction is made between those coming from the Chinese mainland and those coming from other countries. Traditionally, Chinese migrants were allowed free ingress to Hong Kong and they were not subject to immigration controls except during times of emergency. Such a practice was followed partly for diplomatic reasons due to the legal status of the New Territories as an area on lease from China, and partly for practical reasons due to the difficulty of maintaining an impermeable boundary between Hong Kong and the Chinese hinterland. But after the Chinese communist victory in 1949, the policy changed. At that time, a huge flood of refugees poured into Hong Kong. About 1.3 million Chinese entered the territory between 1945 and 1949, doubling the size of the Hong Kong population within a few years. In reaction to this influx, the Hong Kong government closed the border and imposed immigration controls on Chinese travellers in the early 1950s. This did not stop, however, the inflow of Chinese immigrants. For three decades after 1949, the Hong Kong government enacted the so-called 'reach base' policy: that is, if Chinese illegal immigrants were not caught at the border and managed to reach town, they were permitted to stay.

Then in the late 1970s, when China started its economic reform and open-door policies, the influx of Chinese illegal immigrants into Hong Kong reached a crescendo. In 1979, some 89,940 illegal immigrants were arrested upon entry, and a further 102,826 were estimated to have evaded capture. In response to the worsening situation, the Hong Kong government abandoned the 'reach base' policy in 1980 and implemented a strict policy of immediate repatriation of all illegal immigrants no matter where they were caught. Subsequently, illegal immigrants caught engaging in employment were further subject to imprisonment before repatriation.

Apparently, the change in policy towards illegal immigrants from China was enacted only after consultation with the Chinese government. The Hong Kong and Chinese governments further agreed to set a daily quota for legal immigration from the Chinese mainland as described in the previous section. However, the allocation of the immigration quota was entirely in the hands of the Chinese government. There was a four-tier system for screening the applications inside China, from county authorities, municipal authorities, the immigration and emigration bureau, to the public security bureau. In the late 1990s, the deputy public security minister of China announced that a points system would be introduced in the near future to make the allocation more transparent and to avoid possible abuse (South China Morning Post, 6 February 1996). After China resumed sovereignty over Hong Kong in 1997, the existing control over population movement across the border continued to apply. Hong Kong residents were free to move in and out of the mainland, but mainland residents were restricted in their entry into Hong Kong. It is stipulated in the Basic Law that:

> [f]or entry into the Hong Kong Special Administrative Region, people from other parts of China must apply for approval. Among them, the number of persons who enter the Region for the purpose of settlement shall be determined by the competent authorities of the Central People's Government after consulting the government of the Region.

As for non-Chinese immigrants, British nationals enjoyed special privileges and free access to the territory, but this colonial prerogative was abolished after 1997. People from other countries are required to apply for work permits before they can come to live in Hong Kong. There are no quota restrictions imposed on those who come as managers, professionals and other skilled personnel. For unskilled workers, they are brought in under a variety of imported labour schemes.

Malaysia

Malaysia, as a political entity, is geographically divided into east and west Malaysia, separated for approximately 800 miles by the South China Sea. East Malaysia comprises the states of Sabah and Sarawak; while West or Peninsula Malaysia, consists of 12 semi-autonomous states. These states form a federation; a parliamentary democracy headed by a constitutional monarchy.

Malaysia's multi-ethnic population is categorised by the government into two types, viz. the *Bumiputra* and *Non-Bumiputra*. The former term, Bumiputra, means literally the 'sons of the soil'. The term non-Bumiputra refers to those of immigrant descent, chief of which are the Chinese and Indians. According to the last census, carried out in 1991, the population was over 18.55 million, of which 6 percent were non-citizens. Of the 17.5 million Malaysians, 61.7 per cent were Bumiputra, 27.3 percent ethnic Chinese, 7.7 percent ethnic Indians, and the rest falling under the category of 'other'. The majority of the population, i.e., over 80 percent, lived in Peninsular Malaysia. In 1995, the population was estimated at 20 million with a labour force of about eight million.

During the colonial period, Malaysia (then Malaya and British North Borneo) was predominantly an agricultural country dependent for export earnings on two primary commodities: rubber and tin. After independence in 1957, attempts were made by the government to reduce the country's dependence on tin and rubber and to steer the country towards industrialization. To achieve these objectives the government formulated and implemented a series of five-year development plans. Since Malaysia's inception in 1965, six development plans have been implemented: the First Malaysia Plan (1965/70); the Second Malaysia Plan (1971/75); the Third Malaysia Plan (1976/80); the Fourth Malaysia Plan (1981/85); the Fifth Malaysia Plan (1986/90) and the Sixth Malaysia Plan (1991/95). The seventh one is expected to be launched very soon.

The Second Malaysia Plan which encompassed the New Economic Policy (NEP) laid the foundation for industrialization in the country. Two decades after its implementation, manufacturing, palm oil and petroleum replaced rubber and tin as the mainstay of the economy. Manufacturing, for example, accounted for 32.4 percent of GDP in 1995 and 25.5 percent of employment. GDP growth is now 9.5 percent, per capita, with a GDP of USD 8,763 and an unemployment rate of 2.9 percent. The growing economy has its attendant problems. While employment grew at a rate of 3.2 percent per annum between 1990 and 1995, labour supply increased at only 2.9 percent. The resultant labour shortage was made worse by the selective attitude of local labourers due to the expansion of education, improvement in living conditions and access to upward mobility.

Throughout the 1970s no action was taken to address the issue of labour shortage and the infiltration of illegal aliens. It was only when the number of illegals became large and noticeable, and their presence began to cause problems for the local population (especially in the urban areas) that steps were taken to address the problem. The first of such measures was the formation of the *Jawatankuasa Pengambilan Pekerja Asing* (lit. 'Committee for the Recruitment of Foreign Workers'), established in July 1982 and the signing of the Medan Agreement with Indonesia in May 1984. The latter, which was designed to regulate the inflow of Indonesian workers for the plantation sector, was later extended to domestic workers. Subsequently, in 1985, a Memorandum of Understanding (MOU) was signed with the Philippines to import domestic workers. In 1986, permission was given to employers to recruit labour from Bangladesh for the plantation sector, and from Thailand for the plantation and construction sectors. Legal provisions were made to allow the private sector to form agencies for the sole purpose of recruiting alien labour directly from their country of origin.

Without legal avenues for their recruitment, foreign workers in these sectors continued to be recruited illegally. The number of undocumented alien workers in the country continued to grow alongside legally recruited ones and with it the negative consequences of their presence and employment (see below). This caused strong antagonism against the aliens (including legally recruited ones) among some sectors of the public.

An amnesty was first announced in November 1991. Initially it was directed only at domestic workers who were given one month to legalize themselves. The response was encouraging and this induced the government to extend the legislation exercise to the 30 June 1992 to cover those in the construction and plantation sectors. In April 1992, the Immigration Department also registered illegal aliens employed in the manufacturing and service sectors as a result of mounting pressure from manufacturers and in anticipation of an official directive from the government to extend amnesty to them. During the amnesty period, i.e., between November 1991 and 30 June 1992, illegal aliens were to register themselves at thirty registration centres specially set up by the Immigration Department all over the Peninsula.

The registration exercise was accompanied by police measures to prevent further illegal entry, code-named the Ops Nyah I (lit. 'Get Rid Operation I'). Under these on-going operations, the Police Field Force was deployed to patrol over 100 posts, especially along the coasts of Selangor, Negeri Sembilan, Melaka and Johor, where illegal entry by aliens was most active. Under the Ops Nyah II, raids and unannounced checks were carried out in areas suspected of harbouring illegals such as squatter settlements and construction sites. Like the Ops Nyah I, the Ops Nyah II is an on-going exercise.

At the end of 1992, only about 20 percent of those who registered eventually applied for work permits (*The Star*, 26 October 1992). The rest of the registered aliens, approximately 375,000 of them, are technically still illegal as they have not obtained their work permits. In spite of the Ops Nyah I and Ops Nyah II, illegal inflow continues; the crackdown on illegal aliens has not stopped illegal entry by foreigners. Between December 1991 and December 1995, there were, on average, about 53 illegal landings on the Malaysian coasts bringing in on average 1,116 illegal aliens per month. For about the same period, on average 773 illegals were arrested each week under the Ops Nyah II. The limited success achieved by Ops Nyah I and Ops Nyah II, forced the government to look at other ways to curb illegal entry.

Laws and migration-related regulations in Bangladesh

Bangladesh has legislated only the Emigration Ordinance 1982 to regulate the country's labour migration process. Since the 1980s, however, global migration has undergone major changes, and the Ordinance is now less effective in protecting the rights of migrant workers in light of the current global political and economic situation. Many migrants' rights are still violated, and NGOs and advocates believe that the Emigration Ordinance of 1982 needs to be replaced by rights-based legislation reflecting the 1990 UN Convention on the Protection of the Rights of All Migrant Workers and Members of Their Families (MWC) as well as ILO conventions 97 and 143, pertinent to migrant workers. Although the Bangladesh government signed the MWC in 1998, the country has not yet ratified it despite continued calls from migrants' rights advocates and others who have stressed the need for the government to adopt and implement the instruments (Asian Migration Centre 2006).

Bibliography

Abdul-Aziz A.R. 2001. Bangladeshi migrant workers in Malaysia's construction sector: Skills training and language programmes for prospective international workers should be introduced or otherwise expanded. *Asia-Pacific Population Journal*, 16(1), 3-22.

Abella M.I. 1989. Policies and practices to promote remittances. *Philippines Labour Review*, 13(1), 1-17.

Abrar C.R. 2000. *On the margin: Refugees, migrants and minorities*. Dhaka: RMMRU, 222.

Abrar C.R. 2002. Cost-benefit analysis of migration. In *Training manual for community leaders and activists on labour migration process*, edited by Siddiqui Tasneem. Dhaka: RMMRU, 20-24.

Abrar C.R. 2005. How irregularities are committed? *Udbastu*, 33(4).

Acharya S. 2003. *Labour migration in the transnational economies of Southeast Asia*. ESCAP Working Paper on Migration and Urbanization. Bangkok: ESID.

Adams R.H. 1993. The economic and demographic determinants of international migration in rural Egypt. *Journal of Development Studies*, 30(1).

Afsar R. 2001. *Globalization, international migration and the need for networking: The Bangladesh perspective*. Dhaka, Bangladesh Institute of Development Studies.

Afsar R. 2001a. *Internal migration and pro-poor policy*. Dhaka, Bangladesh Institute of Development Studies.

Afsar R. 2003. *Internal migration and the development nexus: The case of Bangladesh*. Regional Conference on Migration, Development and Pro-Poor Policy Choices in Asia, Refugee and Migratory Movements Research Unit and DFID: Dhaka, 22-24 June.

Afsar R., Yunus M. and Islam S. 2000. *Are migrants chasing after the Golden Deer: A study on cost benefit analysis of overseas migration by Bangladeshi Labour*. Geneva: IOM.

Ahlburg D.A. 1995. Migration, remittance and the distribution of income: Evidence from the Pacific. *Asian and Pacific Migration Journal*, 4(1).

Ahmed S.N. 1998. The impact of the Asian crisis on migrant workers: Bangladesh perspectives. *Asian Pacific Migration Journal*, 7(2-3), 369-393.

Ahmed S.R. 2000. *Forlorn migrants: An international legal regime for undocumented migrant workers*. Dhaka: University Press Limited.

Ahmed S.S.U. 2005. Conditions in Bangladesh. In *Migrant workers and the Asian economic crisis: Towards a trade union position*, edited by Holfman Norbert von, Freidrich-Elbert-Stiftung. Office for the Regional Activities in Southeast Asia.

Ainsaar M. 2004. Reason for move: A study on trends and reasons of internal migration with particular interest in Estonia 1989-2000. *Annales Universitatis Turkuensis*. Sarja-ser. bosa-tom. 274. Humaniora: University of Turku.

Akuei S.R. 2005. Remittances as unforeseen burdens: The livelihoods and social obligations of Sudanese refugees. *Global Migration Perspectives*, 18, Geneva, January.

Alam S.M.C. and Rahman M. 2004. *Overview of the Bangladesh population explosion*. [Online] Available at: http://www.power-xs.de/delta/popexpl.html [accessed 14 August 2004].

Alburo F.A. and Abella D. 1992. *The impact of informal remittances of overseas contract workers' earnings on the Philippine economy*. New Delhi: ARTEP.

Alt J. 2005. Life in the world of shadows: The problematic of illegal migration. *Global Migration Perspectives*, 41, Geneva, September.

Andrew G. 2004. Anti-discrimination policy: The emergence of a EU policy paradigm amidst contrasted national models. *West European Politics*, 27(2).

Anthias F. 2000. Metaphors of home: Gendering new migrations to Southern Europe. In *Gender and migration in Southern Europe: Women on the move*, edited by Anthias and Lazaridis. Oxford: Berg Publishers.

Apap J. 2000. Investigating legal labour migration from Maghreb in the nineties. In *Theoretical and methodological issues in migration research: Interdisciplinary, intergenerational and international perspectives*, edited by B. Agozino. Aldershot: Ashgate.

Appleyard R. 1989. International migration and developing countries. In *The impact of international migration on developing countries*, edited by R. Appleyard. France: OECD.

Ariffin R. 2001. *Domestic work and servitude in Malaysia*. Hawke Institute. Working Paper Series, 14. Australia: University of South Australia.

Asato W. 2004. Negotiating spaces in the labour market: Foreign and local domestic workers in Hong Kong. *Asian and Pacific Migration Journal*, 13(2), 255-274.

Asian Development Bank. 2004. *Development outlook 2004*. Bangladesh: ADB.

Asian Migrant Centre. 2000. Asian migrant year book, China, Hong Kong. *Migrant Forum in Asia*.

Asian Migrant Centre. 2004. Asia migration centre year book, Hong Kong. *Migrant Forum in Asia*.

Asian Migrant Centre. 2009. Asian migrant year book, China, Hong Kong. *Migrant Forum in Asia*.

Asian Migrant. 1997. Continuity for migrants in Hong Kong. *Asian Migrant*, 10(3).

Asian Migrant. 2005. Total stock (number) of migrants, as of 2004, Hong Kong. Asian Migrant Centre.

Asis M.B.A. 2000. Imagining the future of migration and families in Asia. *Asian and Pacific Migration Journal*, 9(3), 255-272.

Asis M.B.A., Nicola Piper and Parvati Raghuram. 2010. International migration and development in Asia: Exploring knowledge frameworks. *International Migration*. [Online] Available at: http://www3.interscience.wiley.com/journal/121500407/issue [accessed 10 January 2010]. Published for early view.

Athukorala P.C. 1993. Manufactured exports from developing countries and their terms of trade: A reexamination of the Sarkar-Singer results. *World Development*, 21(10), 1607-1613.

Athukorala P.C. 1993. *Statistics on Asian labour migration: Review of sources, methods and problems*. The Regional Seminar on International Labour Migration Statistics in Asia. New Delhi.

Athukorala P.C. and Manning C. 1999. *Structural change and international migration in East Asia: Adjusting to labour scarcity*. Oxford: Oxford University Press.

Athukorala P.C., Manning C. and Wickramasekara P. 2000. *Growth, employment and migration in Southeast Asia: Structural change in the Greater Mekong Countries*. Cheltenham: Edward Elgar.

Athukorala P.C. and Menon J. 1996. *Export-led industrialization, employment and equity: The Malaysian case*. Research School of Pacific and Asian Studies. Working Paper, 965. Australian National University and Centre of Policy Studies.

Atkinson J. 1990. How gender makes a difference in Wana society. In *How gender makes a difference in Wana society*, edited by Atkinson and Errington, 59-94.

Badawi M. 1989. *Epidemiology of female sexual castration in Cairo, Egypt*. The First International Symposium on Circumcision. Anaheim, California, 1-29 March.

Balbo M. and Marconi G. 2005. Governing international migration in the city of the south. *Global Migration Perspectives*, 38, Geneva, September.

Banerji R. 1993. *Gender gap in Malaysia and Taiwan*. Discussion Paper Series. USA: NORC/University of Chicago, Population Research Center, December.

Bangladesh Bank. 2006. *Report on remittances and foreign exchanges*. Government of the People's Republic of Bangladesh, Dhaka.

Bangladesh Bank. 2007. *Central bank of Bangladesh, Economic data, Country wise remittance inflows*. Dhaka.

Bangladesh Bank. 2010. *Foreign Exchange Policy Department, Bangladesh Bank*. Dhaka.

Barnett C. 1997. Sing along with the common people: Politics, post-colonialism and other figures. *Society and Space*, 15, 137-154.

Battistella G. 2002. Unauthorized migrants as global workers in ASEAN region. *Southeast Asian Studies*, 40(3), 350-371.

Bauböck R. 2005. Citizenship policies: International, state, migrant and democratic perspectives. *Global Migration Perspectives*, 19, Geneva.

Bauer T. and Zimmermann K.F. 1999. *Occupational mobility of ethnic migrants*. IZA Discussion Papers, 58. Germany: Institute for the Study of Labour (IZA).

Baydar N., Michael J.W., Charles S. and Ozer B. 1990. Effects of agricultural development policies on migration in Peninsular Malaysia. *Demography*, 27(1), 97-109.

BBC. 2006. *Migrant workers 'facing problems', UK*. [Online] Available at: http://news.bbc.co.uk/1/hi/scotland/5069778.stm [accessed 11 June 2006].

BBS (Bangladesh Bureau of Statistics). 2001. *Preliminary report of Household Income and Expenditure Survey 2000*. Bangladesh Bureau of Statistics, Dhaka, Ministry of Planning. Government of the People's Republic of Bangladesh.

BBS (Bangladesh Bureau of Statistics). 2002. *Statistical Year Book 1999*. Bangladesh Bureau of Statistics, Ministry of Planning, Dhaka, Government of the People's Republic of Bangladesh.

BBS (Bangladesh Bureau of Statistics). 2004. *Report of the labour force survey 2000*. Bangladesh Bureau of Statistics, Dhaka, Ministry of Planning. Government of the People's Republic of Bangladesh.

BBS (Bangladesh Bureau of Statistics). 2009. *Preliminary estimates of GDP 1999-2000 and final estimates of GDP 1998-1999*. Bangladesh Bureau of Statistics, Dhaka, Ministry of Planning. Government of the People's Republic of Bangladesh.

Beaverstock J.V. and Boardwell J.T. 2000. Negotiating globalization, transnational corporations and global city financial centre in transient migration studies. *Applied Geography*, 20(3), 277-304.

Berninghaus S. and Seifert-Vogt H.G. 1992. Migration and economic development. In Zimmermann K.F. 1992. Munchen: Universitat Munchen.

Bilsborrwo R.E. 1984. Sampling design. In *Migration survey in low income countries: Guidelines for survey and questionnaire design*, edited by R. Bilsborrwo E., Oberai A.S. and Standing G. Kent: Croom Helm.

Black R. 2003. Soaring remittances raise new issues. *Migration Information Source*. Washington: Migration Policy Institute, June.

Blanchet T. 2002. *Beyond boundaries: A critical look at women labour migration and the trafficking within*. Dhaka: USAID.

Blanchet T. 2007. Migration, remittance and the gender gap. Point-counterpoint. *Daily Star*, 5(945), Dhaka, 25 January.

Blau D.M. 1985. The effects of economic development on life cycle wage rates and labour supply behavior in Malaysia. *Journal of Development Economics*, 19(1/2), 163-185.

Blau D.M. 1986. Self-employment, earnings and mobility in Peninsular Malaysia. *World Development*, 18(7), 839-852.

Blayo C. 1989. Problems of measurement. In *The impact of international migration on developing countries*, edited by R. Appleyard. France: OECD.

BMET (Bureau of Manpower Employment and Training). 2002. Recommendations made by the Honourable Advisor of the Ministry of Labour and Employment to reduce the cost of migration and increase wage in a meeting with BAIRA on 22 July 2001, 91-100. In *Beyond the maze: Streaming labour recruitment process in Bangladesh*, edited by T. Siddiqui. Dhaka: RMMRU.

BMET (Bureau of Manpower Employment and Training). 2004. Report. Dhaka: Government of the People's Republic of Bangladesh.

BMET (Bureau of Manpower Employment and Training). 2005. Report. Dhaka: Government of the People's Republic of Bangladesh.

BMET (Bureau of Manpower Employment and Training). 2009. Report. Dhaka: Government of the People's Republic of Bangladesh.

BMET (Bureau of Manpower Employment and Training). 2010. Report. Dhaka: Government of the People's Republic of Bangladesh.

Bohning W.R. and R. Zegers de B. 1995. *The integration of migrant workers in the labour market: Policies and their impact.* International Migration Papers, 8, 1-59. Geneva: International Labour Office.

Borjas G.J. 1994. The economics of immigration. *Journal of Economic Literature*, XXXII, 1667-1717.

Bowring P. 2005. The puzzle of Bangladesh. *The International Herald Tribune*, Hong Kong, 7-8 May.

Boyd M. 1989. Family and personal networks in international migration: Recent developments and new agendas. *International Migration Review*, 23(3), 638-670.

Branden N. 1997. *What self-esteem is and is not.* [Online] Available at: http://www.nathanielbranden.com/catalog/articles_essays/what_self_esteem.html [accessed 3 November 2006].

Bratsberg B. and Dek T. 2002. School quality and returns to education of U.S. immigrants, *Economic Inquiry*, 40(2), April, 177-198.

Brettell C.B. and Hollifield J.F. 2000. Introduction. Migration theory: Talking across discipline. In *Migration theory: Talking across disciplines*, edited by C.B. Brettel and J.F. Hollifield. London: Routledge.

Brettell C.B. and Hollifield J.F. 2000a. Theorizing migration in anthropology: The social construction of networks, identities, communities and globalscapes. In *Migration theory: Talking across disciplines*, edited by C.B. Brettell and J.F. Hollifield. London: Routledge.

Brown P.C. 1992. Migrant remittances, capital flight and macroeconomic imbalance in Sudan's hidden economy. *Journal of African Economics*, 1(1), 86-108.

Brown P.C. 1995. *Consumption and investments from migrants' remittances in the South Pacific.* Geneva: The International Labour Organization.

Brown T.A. 1988. *Migration and politics: The impact of population mobility on American voting behaviour.* USA: The University of North Carolina Press.

Campbell E.H. 2005. Formalizing the informal economy: Somali refugee and migrant trade networks in Nairobi. *Global Migration Perspectives*, 47, Geneva, September.

Capulong A.R.T. 2001. *International laws are no guarantee for migrant workers' rights.* Public Interest Law Center (PILC). The International

Migrants Conference on Forced Labour Export and Forced Migration Amidst Globalization. Philippines, 5 November.

Carballo M. and Nerurkar A. 2001. Migration, refugees and health risks. *Emerging Infectious Diseases*, 7(3), Supplement. International Centre for Migration and Health.

Carl G.H. and Fetzer H. 2001. *The philosophy of Carl G. Hempel: Studies in science, explanation and rationality.* Oxford: Oxford University Press.

Carling J. 2005. *Migrant remittances and development cooperation.* Oslo: International Peach Research Institute, Norway.

Carling J. 2005a. Gender dimensions of international migration. *Global Migration Perspectives*, 35, Geneva.

Casarico A. and Carlo D.C. 2003. Social security and migration with indigenous skill upgrading. *Journal of Public Economics*, 87, 773-797.

Castles S. 1998. *Globalization and migration: Some pressing contradictions.* Oxford: Blackwell Publishers.

Castles S. 1999. How national states respond to immigration and ethnic diversity. In *Migration and social cohesion*, edited by S. Vertovec. London: Edward Elgar.

Castles S. 2003. Migrant settlement, transnational communities and state strategies in the Asian Pacific region. In *Migration in the Asia Pacific: Population, settlement and citizenship issues*, edited by R. Iredale, Hawksley and S. Castles. Cheltenham: Edward Elgar.

Castles S. 2003a. The international politics of forced migration. *Development*, 46(3),11-20.

Castles S. and Millar M.J. 2003. *The age of migration.* 3rd edition. New York: Palgrave Macmillan.

Census and Statistics Department. 2002-2003. *The annual report.* Hong Kong: Hong Kong Government, Chapter 4.

Census and Statistics Department. 2006. *Hong Kong information on immigration.* [Online] Available at: http://www.info.gov.hk/censtatd/hkstat/fas/01c/cd0052001_index.html [accessed 3 November 2006].

Chalamwong Y. 2001. Integration of foreign workers and illegal employment in Thailand. In *International migration in Asia: Trends and policies.* France: OECD.

Chami R.C., Fullenkamp and Jahjah S. 2003. *Are immigrant remittance flows a source of capital for development?* IMF Working Papers, 03/189. Washington: International Monetary Fund.

Chan A., Mary B. and Albert H. 2002. Changes in subjective and objective measures of economic well-being and their interrelationship among the elderly in Singapore and Taiwan. *Social Indicators Research*, 57, 263-300.

Chan R.K.H. and Abdullah M.A. 1999. *Foreign labour in Asia: Issues and challenge.* New York: Nova Science Publishes.

Chantavanich S. 1997. The current situations and impact of labour migration in ASEAN countries. Institute of Asian Studies, Bangkok. *ARCM Newsletter*, 18(12).

Chantavanich S., Germershausen A. and Beesey A. (eds) 2000. *Thai migrant workers in East and Southeast Asia 1996-1997*. Asian Research Centre for Migration, Institute of Asian Studies. Bangkok: Chulalongkorn University, ARCM, 019.

Chantavanich S. and Risser G. 1998. Intra-regional migration in Southeast and East Asia: Theoretical overview, trends of migratory flows and implications for Thailand and Thai migrants workers. In *Thai migrant workers in East and Southeast Asia 1996-1997*, edited by S. Chantavanich, A. Germershausen and A. Beesey, Asian Research Centre for Migration, Institute of Asian Studies. Bangkok: Chulalongkorn University.

Chantavanich S. and Risser G. 2000. Intra-regional migration in Southeast and East Asia: Theoretical overview, trends of migratory flows and implications for Thailand and Thai migrants workers. In *Thai migrant workers in East and Southeast Asia 1996-1997*, edited by S. Chantavanich, A. Germershausen and A. Beesey. Asian Research Centre for Migration, Institute of Asian Studies. Bangkok: Chulalongkorn University.

Chattopaddhyay A. 1995. *Implications of family migration for occupational mobility of women*. The Annual Meeting of the Population Association of America, San Francisco.

Chau N.H.Y. 1997. Migrant networks and the pattern of migration. *Journal of Regional Science*, 37(1), 35-54.

Cheung T. and Mok B.H. 1998. How Filipina maids are treated in Hong Kong – a comparison between Chinese and Western employees. *Social Justice Research*, 11(2), 173-192.

Chiswick B.R. 2000. Are immigrants favourably self selected: An economic analysis? In *Migration theory: Talking across discipline*, edited by C.B. Brettell and J.F. Hollifield. London: Routledge, Chapter 3.

Chitose Y. 2001. The effects of ethnic concentration on internal migration in Peninsular Malaysia, *Asian and Pacific Migration Journal*, 10(2), 241-272.

Chiu S.W.K. 2001. Recent economic and labour migration-related developments in Hong Kong, China. In *International migration in Asia: Trends and policies*. France: OECD, Part 8.

Chiuri M.C., Giuseppe A. and Govianno F. 2005. Crisis in the countries of origin and illegal immigration into Europe via Italy. *Global Migration Perspectives*, 53, Geneva, October.

Chowdhury R.I., Shah N.M., Shah M.A. and Menon I. 2002. Foreign domestic workers in Kuwait: Who employs how many? *Asian and Pacific Migration Journal*, 11(2), 247-270.

Christine B.N.C. 2003. Visible bodies, invisible work: State practices toward migrant women domestic workers in Malaysia. *Asian and Pacific Migration Journal*, 12(1-2), 49-74.

Clark W.A.V. and Onak J.L. 1983. Life cycle and housing adjustment as explanations of residential mobility. *Urban Studies*, 20, 47-57.

Cochran W.G. 1963. *Sampling Techniques*. 2nd edition. New York: John Wiley and Sons.

Cockerham W.C., Abel T. and Lüschen G. 1993. Max Weber, formal rationality and health lifestyles. *The Sociological Quarterly*, 34(3), 413-428.

Cohen R. 1996. *Sociology of migration*. London: Edward Elgar.

Cohen R. 1996a. *Theories of migration*. London: Edward Elgar.

Coleman J. 1973. *The mathematics of collective action*. London: Heinemann.

Collins A. 2004. Sexuality and sexual services in the urban economy and socialscape: An overview. *Urban Studies*, 41(9), 1631-1641.

Connell J. and Brown R. 1995. Migration and remittance in the South Pacific: Towards new perspective. *Asian and Pacific Migration Journal*, 4(1).

Cox D. 1997. The vulnerability of Asian women migrant workers to a lack of protection and to violence. *Asian and Pacific Migration Journal*, 6(1), 59-75.

Cullinane S. and Cullinane K. 2003. City profile: Hong Kong. *Cities*, 20(4), 279-288. Amsterdam: Elseveier.

Cwerner S.B. 2001. The times of migration. *Journal of Ethnic and Migration Studies*, 27(1), 7-36.

Daily Ajker Kagoj. 2004. 27.5 million unemployed people are looking for jobs. Dhaka, 26 December.

Daily Ajker Kagoj. 2004. Attempt to board with fake passport and visa: Arrested 450 between 2001 and 2004 at Zia International Airport. Dhaka, 16 July.

Daily Inqilab. 2004. More than 10 thousands of Bangladeshis in Malaysian prison. Dhaka, 10 July.

Daily Inqilab. 2005. Young were kept captive 8 days in the desert without any food. Dhaka, 28 March.

Daily Ittefaq. 2005. International terrorists hold Bangladesh passports. Dhaka, 16 July.

Daily Ittefaq. 2007. Bangladeshi workers abroad have been remaining without any job. Dhaka, 22 January.

Daily Jugantor. 2004. Illegal workers in Malaysia to get legal status. Dhaka, 7 July.

Daily Prothom Alo. 2004. Disasters are on increase in Bangladesh. Dhaka, 19 August.

Daily Prothom Alo. 2005. Manpower export: Many victims losing everything falling prey of the fraud barkers. Dhaka, 8 July.

Daily Star. 2002. Huge job scopes in Malaysia for Bangladeshi workers, 3(969), 1 June.

Daily Star. 2004. Foreign workers in Saudi systematically abused: Some even treated like slaves: Rights group. Dhaka, 16 July.

Daily Star. 2005. 4.37 lakh female migrants working abroad, 5(485). Dhaka, 6 October.

Daily Star. 2006. 31 women held at ZIA on way to Libya with fake papers, 5(880). Dhaka, 18 November.

Daily Star. 2010. Expatriate welfare bank pledged. Dhaka.

Dannecker P. 1999. *Conformity or resistance: Women workers in the garment factories in Bangladesh*. Working Paper, 326. Sociology of Development Research Centre, University of Bielefeld.

Dannecker P. 2003. *The construction of the myth of Malaysia: Labour migration from Bangladesh to Malaysia*. University of Belfield Working Paper, 345. Bielefeld.

Dannecker P. 2005. Transnational migration and the transformation of gender relations: The case of Bangladeshi labour migrants. *Current Sociology*, 53(4), 655-674.

Dannecker P. 2006. *Bangladesh: Double standards. Magazine for development and cooperation. Women Living Under Muslim Laws (WLUML)*. [Online] Available at: http://www.wluml.org/english/newsfulltxt.shtml?cmd[157]=x-157-536525 [accessed 15 August 2006].

DaVanzo J. 1982. Techniques for analysis of migration-history data. In *Economic and social commission for Asia and the Pacific*, National Migration Surveys: X. Issued as P-6760, Techniques for Analysis of Migration-History Data from the ESCAP National Migration Surveys, RAND, May.

Davidmann M. 2006. *Work and pay, incomes and differentials: Employer, employee and community*. [Online] Available at: http://www.solbaram.org/articles/clm4.html.

Davies R. 2000. Neither here nor there? The implications of global diasporas for (inter)national security. In *Migration, globalisation and human security*, edited by D.T. Graham and N.K. Poku. London: Routledge.

DeFay. 2004. *The sociology of international migration*. California: University of California.

DeJong G.F. and Fawcett J.T. 1981. Motivation for migration: An assessment and a value expectancy model. In *Migration decision making: Multidisciplinary approaches to micro level studies in developed and developing countries*, edited by DeJoung G.F. and Gardner R.W. New York: Pergamon Press.

Demleitner N.V. 2001. *The law at crossroads: The construction of migrant women trafficked into prostitution*. Baltimore: Johns Hopkins University Press.

Demuth A. 2000. Some conceptual thoughts on migration research. In *Theoretical and methodological issues in migration research: Interdisciplinary, inter-generational and international perspectives*, edited by B. Agozino. Farnham: Ashagte.

DeVoretz J. 2004. Immigration policy: Methods of economic assessment. *Global Migration Perspectives*, 4, Geneva.

Dictionary of the English Language. 2004. Dictionary definition of promise. The American Heritage. Houghton Mifflin Company.

Djajic S. and Milbourne R. 1988. A general equilibrium model of guestworker migration. *Journal of International Economics*, 25, 335-351.

Doorn J. Van. 2000. *Migration, remittance and small enterprise development.* Geneva: ILO.

Dorall F.R. 1989. *Foreign workers in Malaysia: Issues and implications of recent illegal economic migration from the Malay world? The trade in domestic helpers.* Kuala Lumpur: PDC, 287-16.

Doyle M.W. 2004. The challenge of worldwide migration. *Journal of International Affairs,* 57(2), 1-5.

Dudley L.P. 2002. Human capital, cultural capital and economic attainment patterns of Asian-born immigrants to the United States: Multi-level Analyses. *Asian and Pacific Migration Journal,* 11(2),197-220.

Dustmann C. 1993. Earnings adjustment of temporary migrants. *Journal of Population Economics,* 6, 153-168.

Dustmann C. 1994. Speaking fluency, writing fluency and earnings of migrants. *Journal of Population Economics,* 7, 133-156.

Dustmann C. 1997. Return migration, uncertainty and precautionary savings. *Journal of Development Economics,* 52, 295-316.

Eagly H.A. and Shelly C. 1993. *Process theories of attitude formation and change: Reception and cognitive responding. The psychology of attitudes.* San Diego, CA: Harcourt Brace Jovanocvich Publishers.

El-Sakka M.I.T. and McNabb R. 1999. The macroeconomic determinants of emigrant remittance. *World Development,* 27(8), 1493-1502.

Elster J. 2000. *Strong feelings emotion, addiction and human behavior.* Cambridge: MIT Press.

Elwell F. 1999. *Cultural materialism: A sociological revision.* Westport Connecticut: Praeger Press.

Emerton R. and Peterson C. 2003. *Migrant nightclub/escort workers in Hong Kong: An analysis of possible human rights violations.* Occasional Paper, 3. Faculty of Law, University of Hong Kong.

Esser H. 2003. *Does the new immigrant require a new theory of intergenerational integration?* The Centre for Migration and Development, CMD Working Paper, 03-09k. Princeton.

Faini R. and Venturini A. 2001. *Home bias and migration: Why is migration playing a marginal role in the globalization process?* Working Paper, 27. Centre for Household, Income, Labour and Demographic Economics (CHILD).

Faist T. 2000. *Trans-nationalization in international migration: Implications for the study of citizenship and culture.* Institute for Intercultural and International Studies (InIIS), WPTC-99-08, Bremen.

Family Income and Expenditure Survey (FIES). 2000. *Final version of the annual report.* The Philippines: NSO Publications.

FAO (Food and Agriculture Organization). 1983. *World Programme for the Census of Agriculture 2010.* Community level data, Chapter 5. Geneva.

Fei G. 2002. School attendance of migrant children in Beijing, China: A multivariate analysis. *Asian and Pacific Migration Journal,* 11(3), 357-374.

Findlay A.F., Li L.N., Jowettt A.J. and Ronald S. 1996. Skilled international migration and the global city: A study of expatriates in Hong Kong. *Trans Inst Br Geogr*, 21, 49-61.

Forest Department. 2005. [Online] Available at: http://www.forest.go.th/eng [accessed 22 June 2005].

Forrest J., Pulsen M. and Johnston R. 2003. Everywhere different? Globalization and the impact of international migration on Sydney and Melbourne. *Geoforum*, 34, 499-510.

Friedberg A.L. 2000. *The shadow of the Garrison State: America's anti-statism and its Cold War strategy*. Princeton, NJ: Princeton University Press.

Friedrichs J. 2002. Globalization, urban restructuring and employment prospects: The case of Germany. In *Globalization and the new city: Migrants, minorities and urban transformations in comparative perspective*, edited by M. Cross and R. Moore. London: Palgrave.

Gammeltoft P. 2002. *Remittance and other financial flows to developing countries*. CDR Working Paper, 2, 11. Copenhagen.

Gaude J. and Peek P. 1976. The economic effect of rural urban migration. *International Labour Review*, 111(3), 333-8.

Gazi R., Chowdhury Z.H., Alam S.M.N., Chowdhury E., Ahmed F., Begum S. 2001. *Trafficking of women and children in Bangladesh: An overview*. Dhaka, ICDDR,B: Centre for Health and Population Research. Special publication, 111, 18.

Ghai D. 2004. Diasporas and development: The case of Kenya. *Global Migration Perspectives*, 10, Geneva.

Ghatak S. 1995. *Introduction to development economics*. London: Routledge.

Ghosh S. 1988. *An ordered probit model of the determinants of the type of inter-household transfer: Some evidence from the Malaysian family life survey*. Michigan: University of Michigan.

Gillan M. 2002. Refugees or infiltrators? The Bharatiya Janata Party and 'illegal' migration from Bangladesh. *Asian Studies Review*, 29(1), 73-95.

Glavaz S.M. and Waldrof B. 1998. Segregation and residential mobility of Vietnamese immigrants in Brisbane, Australia. *Professional Geographer*, 50(3), 344-357.

Glick C.E. 1968. *The changing positions of two Tamil groups in Malaysia 'Indian' Tamils and 'Ceylon' Tamils*. Second International Conference Seminar of Tamil Studies. Chennai: Tamil Nadu.

GoB (Government of Bangladesh). 1982. *The emigration ordinance 1982. Dhaka Law Report*. Dhaka.

GoB (Government of Bangladesh). 1998. *Labour force survey*. Dhaka, Bangladesh Bureau of Statistics.

GoB (Government of Bangladesh). 1999. *Report on population growth*. Dhaka, Ministry of Panning.

GoB (Government of Bangladesh). 2000. *Report on the second national expanded HIV surveillance, 1999-2000*. Dhaka, Bangladesh, AIDS and STD Control Programme.

GoB (Government of Bangladesh). 2001. *Population census*. Dhaka, Bangladesh.

GoB (Government of Bangladesh). 2005. *Annual report – 2002*. Dhaka, Ministry of Expatriate Welfare and Overseas Employment.

GoB (Government of Bangladesh). 2006. *Statistical pocketbook: Bangladesh 2001*. Dhaka, Bangladesh Bureau of Statistics.

GoB (Government of Bangladesh). 2008. *Population data sheet*. Dhaka, Ministry of Planning.

GoB (Government of Bangladesh). 2009. *Population census 2001*. Dhaka, Ministry of Panning.

Goldscheider C. 1996. Migration and social structure: Analytic issues and comparative perspective in developed nations. In *The sociology of migration*, edited by R. Cohen. London: Edward Elgar.

Golledge R. and Stimson R.J. 1997. *Spatial behavior: A geographic perspective*. New York: The Guilford Press.

Green A. 1995. The geography of dual career households: A research agenda and selected evidence from secondary data sources for Britain. *International Journal of Population Geography*, 1, 29-50.

Grieco E.M. and Boyd M. 2003. *Women and migration: Incorporating gender into international migration theory*. Centre for the Study of Population. Working Paper, 98-139. Florida Sate University.

Guerrero G.T. and Bolay J. November 2005. Enhancing development through knowledge circulation: A different view of the migration of highly skilled Mexicans. *Global Migration Perspectives*, 51, Geneva.

Gunatilleke G. (ed.) 1986. *Migration of Asian workers to the Arab world*. Tokyo: United Nations University Press.

Gurowitz A. 2000. Migrant rights and activism in Malaysia: Opportunities and constraints. *The Journal of Asian Studies*, 59(4), 863-888.

Haas H. 2005. International migration, remittances and development: Myths and fact. *Global Migration Perspectives*, 30, Geneva.

Habermas J. 1984. *The theory of communicative action. Reason and the rationalization of society* [Trans. T. McCarthy]. Cambridge: Polity Press.

Haider M. 2005. Factors that deprived Bangladesh of the opportunity of labour export to Malaysia. *The Naba Diganta*. Dhaka, 5 May.

Hamilton A. 2002. Tribal peoples on the southern Thai border: Internal colonialism, minorities and the State. In *Tribal communities in the Malay world*, edited by B. Geoffrey and C. Cynthia. Singapore: ISEAS/Leiden: IIAS, 77-96.

Haque S. 2004. Foreword. In *Revisiting the human trafficking paradigm: Bangladesh experiences*. Geneva: IOM.

Harcourt W. 2003. The changing face of migration. *Development*, 46(3), 3-5.

Harris J.R. and Todaro M.P. 1970. Migration, unemployment and development: A two-sector analysis. *The Amer. Econ. Review*, 60(1), 126-38.

Hasan R. 2002. Recruiters rip recruits off: Workers pay 'unofficial fees' at several stages thru' middlemen. *Daily Star*, 3(986). Dhaka, 18 June.

Hasan R. 2005. Move to close down illegal travel agencies. *Daily Star*. Dhaka, 24 March.

Hassan M. (n.d.). *Complementarities between international migration and trade: A case study of Bangladesh*. Southwestern College, Kansas.

Hawwa S. 2000. From cross to crescent: Religious conversion of Filipina domestic helpers in Hong Kong. *Islam and Christian-Muslim Relation*, 11(3), 347-67.

Heath A. 1976. *Rational choice and social exchange*. Cambridge: Cambridge University Press.

Hefti A.N. 1997. *Globalization and migration. The European solidarity conference on the Philippines: Responding to globalization*. Boldern House, Männedorf: Zurich, 19-21 September.

Heisler B.S. 1999. The future of immigrant incorporation: Which models? Which concepts? In *Migration and social cohesion*, edited by S. Vertovec. London: Edward Elgar.

Hewison K. 1999. *Localism in Thailand: A study of globalization and its discontents. Centre for the study of globalization and regionalization*. Working Paper, 39/99. University of Warwick.

Hewison K. 2003. *A preliminary analysis of Thai workers in Hong Kong: Survey results*. Southeast Asian Research Centre and CAPSTRANS/City University of Hong Kong. Working Paper, 44. Hong Kong.

Hewison K. and Young K. (ed.) 2006. Transnational migration and work in Asia. Routledge/City University of Hong Kong Southeast Asian Studies.

Ho L., Lui P. and Lam K. 1991. *International labour migration: The case of Hong Kong*. Hong Kong Institute of Asia-Pacific Studies. The Chinese University of Hong Kong: Hong Kong. Occasional Paper, 8.

Hollifield, J. 2000. The politics of international migration: How can we bring the State back. In *Migration theory: Talking across disciplines*, edited by C.B. Brettel and J.F. Hollifield. New York and London: Routledge, 137-85.

Hossain M.A. and Ullah A.K.M.A. 2004. Urban governance in Bangladesh: Exploring contemporary issues. *BIISS Journal*, 25(2).

Hugo G. 1998. The demographic underpinnings of current and future international migration in Asia. *Asian Pacific Migration Journal*, 7(1), 1-25.

Huguet J.W. 1989. International labour migration from the ESCAP region. In *The impact of international migration on developing countries*, edited by R. Appleyard. France: OECD.

Hye H.A. 1996. *Below the line: Rural poverty in Bangladesh*. Dhaka: University Press Limited, Chapter IV, 112.

ILO (International Labour Organization). 1994-1995. *Statistics on international labour migration: A review of sources and methodological issues*. Inter-departmental project on migrant workers. Geneva.

ILO (International Labour Organization). 1995. *Migration from the Maghreb and migration pressures: Current situation and future prospects*. Geneva.

ILO (International Labour Organization). 1999. *Decent work.* Report of the Director General, International Labour Conference, 87th session, Geneva. Working Paper, 66.

ILO (International Labour Organization). 2000. *Making the best of globalization.* Concept Paper, Workshop on Making the Best of Globalization: Migrant Worker Remittances and Microfinance. Geneva: ILO.

ILO (International Labour Organization). 2001. *Reducing the decent work deficit: A global challenge.* Geneva.

ILO (International Labour Organization). 2002. *Migrant workers, labour education 2002/4.* No. 129, Geneva.

ILO (International Labour Organization). 2003. *Decent work in Denmark: Employment, social efficiency and economic security.* Geneva.

ILO (International Labour Organization). 2004. *Facts on migrant labour.* Switzerland.

IMF (International Monetary Fund). 1995. *International financial statistics yearbook – 1995 and Malaysia.* Ministry of Finance, Economic Report 1995 (for estimates for 1995).

IMF (International Monetary Fund). 2002. *International financial statistics.* Washington.

IMF (International Monetary Fund). 2005. *Balance of payments statistics.* Washington.

Immigration Department. 2002-2003. *Annual report,* Chapter 7. Hong Kong.

Independent, The. 2005. Government to take stern action illegal manpower exporters, 24 March.

Independent, The. 2005. Middle East miffed at Bangladeshi workers. Dhaka, 23 March.

INSTRAW/IOM. 2000. *Temporary labour migration of women: Case studies of Bangladesh and Sri Lanka.* United Nations International Research and Training Institute for the Advancement of Women (INSTRAW) and International Organization for Migration (IOM), Dominican Republic.

IOM (International Organization for Migration) and UNDP. 2002. *Proceedings of national consultation workshop on labour migration process in Bangladesh.* Dhaka.

IOM (International Organization for Migration). 2000. *World migration report 2000.* Geneva.

IOM (International Organization for Migration). 2003. Defining migration priorities in an interdependent world, *Migration Policy Issues,* 1, 1-6.

IOM (International Organization for Migration). 2003. *World migration report 2003.* Geneva.

IOM (International Organization for Migration). 2005. Dynamics of remittance utilization in Bangladesh, *IOM Migration Research Series,* 18, Geneva, 1-95.

IOM (International Organization for Migration). 2008. *World migration 2008. Managing labour mobility in the evolving global economy.* Geneva.

IOM (International Organization for Migration). IOM. 2003. Defining migration priorities in an interdependent world. *Migration Policy Issues*, 1, 1-6.

IOM (International Organization for Migration). September 2004. *Revisiting the human trafficking paradigm: Bangladesh experiences*. Geneva.

IOM (International Organization for Migration). 2009. *Migration in South East Asia*. [Online] Available at: http://www.iom-seasia.org/ [accessed 1 January 2010].

Iredale R., Hawksley C. and Castles S. 2003. Foreword. In *Migration in the Asia Pacific: Population, settlement and citizenship issues*, edited by R. Iredale, C. Hawksley and S. Castles. London: Edward Elgar.

Islam M.S., Mazharul H., Delwar H. and Kazi S.K. 2006. The unheard voices of 'Orang Bangla'. In *Malaysia*. [Online] Available at: Countercurrents.org [accessed 4 January 2010].

Islam T. 2005. Deaths in Chasing the Golden Deer. *Udbastu*, 33, 1. Dhaka, July-September.

Israel G.D. 1992. *Determining sample size*. Programme evaluation and organizational development, IFAS, University of Florida, PEOD-6, November.

Iyer A., Theresa W.D. and Brenda S.A.Y. 2004. A clean bill of health: Filipinas as domestic workers in Singapore. *Asian and Pacific Migration Journal*, 13(1), 11-38.

Jackson J.A. 1986. *Aspects of modern sociology: Migration*. New York: Longman.

Jaijaidin. 2006. *Lasher sobi: E abar emon ki?* [Picture of dead bodies: Does it matter at all?], 22(29), 2 May.

Jamil K. and Rebecca W. 1993. *Income aspirations and migrant women's labour force activity in Malaysia*. Johns Hopkins Population Center Papers on Population, WP 93-04, Johns Hopkins University.

Jandl M. 2005. The Development-visa scheme: A proposal for a market-based migration control policy. *Global Migration Perspectives*, 36, Geneva.

Jones S. 1996. Hope and tragedy for migrants in Malaysia. *Asia-Pacific Magazine*, 1, 23-27, April.

Kabeer N. 1997. Women, wages and intra-household power relations in urban Bangladesh. *Development and Change*, 28, 261-302.

Kabeer N. 2003. *Gender mainstreaming in poverty eradication and the millennium development goals: A handbook for policy-makers and other stakeholders: Gender management system series*. International Development Research Centre, Ottawa.

Kabeer N. 2004. Globalization, labour standards and women's rights: Dilemmas of collective (in)action in an interdependent world. *Feminist Economics*, 10(1), 3-35.

Kahanec M. and Zimmermann K.F. 2008. *Migration and globalization: Challenges and perspectives for research infrastructure*. IZA DP, 3890. Germany: IZA.

KAKAMNP. 1998. *Migrant workers and Asian economic crises: Towards a trade union position*. The Southeast Asian Regional Conference on Migrant Workers, Bangkok, 5-6 November.

Kam-yee L. and Lee K. 2006. Citizenship, economy and social exclusion of Mainland Chinese immigrants in Hong Kong. *Journal of Contemporary Asia*, 36(2).

Kanapathy V. 2001. International migration and labour market adjustments in Malaysia: The role of foreign labour management policies. *Asian Pacific Migration Journal*, 10(3-4), 429-461.

Kapur D. 2004. Ideas and economic reforms in India: The role of international migration and the Indian diaspora. *India Review*, 3(4), 364-384.

Kardar S. 1992. *Exchange and payment reforms in Pakistan*. New Delhi: UNDP/ ILO-ARTEP.

Karim M.F. 2006. Bangladesh manpower export is a neglected industry. *The Bangladesh Observer*. Dhaka, 9 May.

Karn S. 2006. *Migration in Rural Nepal*. SAMRen Paper. Dhaka, June.

Kassim A. 2001. Integration of foreign workers and illegal employment in Malaysia. In *International migration in Asia: Trends and policies*. France: OECD, Chapter XII.

Kassim A. 2001a. Recent trends in migration movements and policies in Malaysia. In *International migration in Asia: Trends and policies*. France: OECD.

Kazi S. 1989. Domestic impact of overseas migration: Pakistan. In *To the Gulf and back: Studies on the economic impact of Asian labour migration*, edited by R. Amjad. New Delhi: ILO/ARTEP, 167-96.

Kephart J.L. 2005. *Immigration and terrorism moving beyond the 9/11 staff report on terrorist travel*. Washington: Center for Immigration Studies, September.

Keynes J.M. 1935. *The general theory of employment, interest and money*. Cambridge: King's College.

Khalaf S. and Alkobaisi S. 1999. Migrants' strategies of coping and patterns of accommodation in the oil-rich Gulf societies: Evidence from the UAE. *British Journal of Middle Eastern Studies*, 26(2), 271-298.

Khan M.G.A. 2003. Cases of 98 Bangladeshi dead still on ice. *Riyadh's Arab News*, 7 August.

Khan M.M.R. 2005. Child trafficking. *The Independent*. Dhaka, 18 March.

Khan S.A. 2004. One hundred thousand labours are being sent back from Malaysia. *Daily Inqilab*. Dhaka, 28 July.

Khanum S.M. 2001. The household patterns of a Bangladeshi village in England. *Journal of Ethnic and Migration Studies*, 27(3), 489-504.

Kibria N. 2004. *Returning international labour migrants from Bangladesh: The experience and effects of deportation*. Working Paper, 28. University of Boston, Department of Sociology.

Kim W.B. 1996. Economic interdependence and migration dynamics in Asia. *Asian Pacific Migration Journal*, 5(2-3), 303-317.

Kofman E. 2004. Gendered global migrations: Diversity and stratifications. *International Feminist Journal of Politics*, 6(4), 643-665.

Krieger H. 2004. *Migration trends in an enlarged Europe*. European Foundation for the Improvement of Living and Working Conditions. Dublin: Quality of Life in Central and Eastern European Candidate Countries, UK.

Kritz M.M., Lim L.L. and Zlotnik H. 1992. Global interactions: Migration systems, processes and policies. In *International migration systems: A global approach*, edited by M.M. Kritz, L.L. Lim and H. Zlotnik. Oxford: Clarendon Press.

Kwong P.C.K. 1990. Migration and manpower shortage? In *The other Hong Kong report*, edited by Y.C. Richard and Y.S. Joseph. Hong Kong: Chinese University Press, 297/37.

Kyokai N.R. 2003. *Migration and labour market in Asia: Recent trends and policies*. France: OECD.

Landau L. 2004. Democracy and discrimination: Black African migrants in South Africa. *Global Migration Perspectives*, 5, Geneva, October.

Lariosa J.G. 2006. Money remittances, *The Manila Mail*, XVI, 3.

Law L. 2002. Defying disappearance: Cosmopolitan public spaces in Hong Kong. *Urban Studies*, 39(9), 1625-1645.

Lee E.S. 1966. A theory of migration. *Demography*, 3(1), 47-57.

Lee Y. 2006. Transit network sensitivity analysis. *Journal of Public Transportation*, 9(1), 91-122.

Levitt P. and Nyborg-Sorensen N. 2004. The transnational turn in migration studies. *Global Migration Perspectives*, 6, Geneva.

Lewellen T.C. 2002. *The anthropology of globalization: Cultural anthropology enters the 21st century*. Westport, Conn: Bergin and Garvey.

Li F.L.N., Findlay A.M. and Jones H. 1998. A cultural economy perspective on service sector migration in the global city: The case of Hong Kong. *International Migration*, 36(2).

Light I., Parminder B. and Stavros K. 1990. *Migration networks and immigrant entrepreneurship*. Institute for Social Science Research, V. University of California.

Lindquist B.A. 1993. Migration networks: A case study in the Philippines. *Asian Pacific Migration Journal*, 2(1), 75-104.

LOC. 2004. *Country studies: Bangladesh migration*. [Online] Available at: www.countrystudies.us/bangladesh/27.htm [accessed 18 December 2004].

Logan J., Alba R. and Zhang W. 2002. Immigrant enclaves and ethnic communities in New York and Los Angeles. *American Sociological Review*, 67, 299-322.

Lohrmann R. 1989. Irregular migration: An emerging issue in developing countries. In *The impact of international migration on developing countries*, edited by R. Appleyard. France: OECD.

Lomnitz L. 1977. *Networks and marginality*. New York: Academic Press.

Lowell L. and Kemper Y.B. 2004. Transatlantic roundtable on low-skilled migration in the twenty-first century: Prospect and policies. *International Migration*, 42(1), 118-41.

Lucas R.E.B. 2001. *Diaspora and development: Highly skilled migrants from East Asia*. Boston University Institute for Development Discussion Paper, 120.
Lucas R.E.B. and Stark O. 1985. Motivations to remit: Evidence from Botswana. *Journal of Political Economy*, 93, 901-18.
MacPherson D.W. and Brian D. Gushulak. 2004. Irregular migration and health. *Global Migration Perspectives*, 7, Geneva, October.
Maharaj B. 2004. Immigration to post-apartheid South Africa. *Global Migration Perspectives*, 1, Geneva, August.
Mahmood R.A. 1996. *Immigration dynamics in Bangladesh: Level, pattern and implications*. Dhaka, Asiatic Society of Bangladesh.
Mahmood R.A. 1998. *Globalization, international migration and human development: Linkage and implications*. UNDP (Unpublished).
Malaysiakini. 2006. *Government bans Bangladeshi workers again*. [Online] Available at: http://www.malaysiakini.com/news/57803 [accessed 6 October 2006].
Malynovska O. 2004. International migration in contemporary Ukraine: Trends and policy. *Global Migration Perspectives*, 14, Geneva.
March J.G. and Johan P. Olsen. 2004. *The logic of Appropriateness*. Stanford: Stanford University Arena – Centre for European Studies. Working Papers, 04/09. University of Oslo.
Mardzoeki F. 2002. Women's solidarity for human rights. *Journal of Indonesia Today*. Jakarta, 13 September.
Marques H. 2005. *Migration creation and diversion in the EU: Are CEECs immigrants crowding-out the rest?* Discussion Paper Series 2005. Loughborough: Department of Economics, Loughborough University.
Martin P. and Houstoun M.F. 2001. The future of international labour migration. *Journal of International Affairs*, 311-333.
Martin S.F. 2001. *Global migration trends and asylum. New issues in refugee research*. Working Paper, 41. Washington: Washington, Georgetown University.
Massey D. 1988. Economic development and international migration in comparative perspective. *Population and Development Review*, 14, 383-413.
Massey D. 1989. *Social structure, household strategies and the cumulative causation of migration*. The Annual Meeting of the American Sociological Association. San Francisco Hilton Hotel, 13 August.
Massey D. and Arango J. and Hugo G. 1996. Theories of international migration: A review and appraisal. In *Theories of migration*, edited by R. Cohen. London: Edward Elgar.
Massey D., Arango J., Hugo G., Kouaouci A., Pallegrino A. and Taylor J.E. 1996. Theories of international migration: A review and appraisal. In *Theories of migration*, R. Cohen. Cheltenham: Edward Elgar.
Massey D. and Zenteno. 1999. The dynamics of mass migration. *Proc. Natl Acad. Sci*, 96, 5328-5335.

Menon R. 1987. Job Transfers: A neglected aspect of migration in Malaysia. *International Migration Review*, 21(1), 86-95.

Menon R. 1988. How Malaysian migrants prearrange employment. *Sociology and Social Research*, 73(4), 257-259.

Merkle L. and Zimmermann K.F. 1992. Savings, remittance and return migration. *Economic Letters*, 38 (January), 7-81.

Mian A.Q. 1999. *Applied statistics*. Asian Institute of Technology (AIT). Thailand.

Miaoulis G. and Michener R.D. 1976. *An Introduction to sampling*. Dubuque, Iowa: Kendall/Hunt Publishing Company.

Middleton D. May 2005. Why asylum seekers seek refuge in particular destination countries: An exploration of key determinants. *Global Migration Perspectives*, 34, Geneva.

Migration News. 1997. Malaysia: Roundups and new migrants. *Migration News*, 4(4), May.

Migration News. 2005. Malaysia. *Migration News*, 12(2), April.

Ministry of Human Resources. 2009. *Malaysia, Indonesia to review foreign workers agreement*. Putrajaya: Malaysia.

Miyan A.M. 2003. *Dynamics of labour migration-Bangladesh context*. Dhaka, International University of Business Agriculture and Technology.

Mujeri M. and Khandker B. 2002. *Decomposing wage inequality change in Bangladesh: An application of double calibration technique*. Globalization and poverty project. DFID. UK.

Muller G.O.W. (ed.) 1996. *The general report*. Standing Rapporteur, International conference on migration crime, Courmayer, Italy, 7-9 October.

Münz R. 2004. Migrants, labour markets and integration in Europe: A comparative analysis. *Global Migration Perspectives*, 16, Geneva, October.

Murshid K.A.S., Iqbal K. and Ahmed M. 2002. *A study on remittance inflows and utilization*. Dhaka: IOM, Regional Office for South Asia.

Myles J. and Hou F. 2003. *Changing colours: Spatial assimilation theory and new racial minority immigrants*. Canada: University of Toronto and Statistics Canada.

Nasra S.M. and Menon I. 2002. Factors contributing to the vulnerability of migrant workers in the Gulf. *Udbastu*, 19 and 20. Dhaka, January-June.

Nasra S.M. 1999. Emigration dynamics in South Asia: An overview. In *Emigration dynamics in developing countries, II: South Asia*, edited by R. Appleyard. Aldershot: Ashgate Publishing, 17-29.

Nasra S.M. and Menon I. 1999. Chain migration through social network: Experience of labour migrants in Kuwait. *International Migration*, 37(2), 361-82.

Neft N. and Levine A. 1997. *Where women stand: An international report of the status of women in 140 countries 1997-1998*. New York: Random House.

Netto A. 2001. Kuala Lumpur under fire for ban on Bangladeshis. *Asia Times*, 27 February.

Neumann I.B. 1992. *Regions in international relations theory: The case for a region-building approach.* Norwegian Institute of International Affairs. Research Report, 162. Olso.

New Age, The. 2005. Dreams die in the Mediterranean sea, Dhaka, 11 March.

New Strait Times. 2005. Missing students: College faces closure, Malaysia, 14 May.

Nikolinakos M. 1996. Notes towards a general theory of migration in late capitalist. In *The theories of migration*, edited by R. Cohen. London: Edward Elgar.

Nussbaum M. et al. 2003. *Essays on gender and governance.* Human Development Resource Centre, New Delhi: United Nations development Programme.

Ofreneo R.P. 2000. *Women in Asia and the pacific: A trade union perspective.* Background report. Regional Women's Conference, Kota Kinabalu, Sabah: Malaysia. September.

Oishi N. 2002. *Gender and migration: An integrative approach.* University of California, Working Paper, 49.

Oishi N. March-April 2006. Women in motion: Globalization, state policies and labour migration in Asia. *Canadian Journal of Sociology Online.*

Olsson, G. 1965. Distance and human interaction: A migration study. *Geografiska Annaler*, 47B, 3-43.

Ong J.H., Chan K.B. and Chew S.B. 1995. Citizens and foreign labour in Singapore. In *Crossing borders: Transmigration in Asia Pacific.* Singapore: Simon and Schuster/Prentice Hall.

Ortiz V. 1992. *Circular migration and employment among Puerto Rican women.* University of California. Center of American Politics and Public Policy, Occasional Paper Series, 92-2.

Ossman S. 2004. Studies in serial migration. *International Migration*, 42(4),111-121.

Palloni A. 2001. Social capital and international migration: A test using information on family networks. *American Journal of Sociology*, 106(5), 1262-98.

Panda S.M. 2006. *Women's collective action and sustainable water management: Case of SEWA's water campaign in Gujarat.* IFPRI Working Paper Series, 61.

Papadopoulou A. 2005. Exploring the asylum-migration nexus: A case study of transit migrants in Europe. *Global Migration Perspectives*, 23, Geneva, January.

Parsons T. and Edward S. 2001. Values, Motives and Systems of Actions. In *Toward a general theory of action*, edited by P. Talcott and S. Edward. Boston: Harvard University Press.

Pasha N. 2004. Ten thousand Bangladeshi in Malaysian prison. *Daily Inqilab.* Dhaka, 10 July.

Pécoud A. and Guchteneire P. 2004. Migration, human rights and the United Nations. *Global Migration Perspectives*, 3, Geneva, September.

Pécoud A. and Guchteneire P. 2005. Migration without borders: An investigation into the free movement of people. *Global Migration Perspectives*, 27, Geneva, April.

Pillai P. 1996. Labour market developments and international migration in Malaysia. In *OECD documents: Migration and the labour market in Asia-prospects to the year 2000*. France: OECD.

Pillai P. 1998. The impact of the economic crisis on migrant labour in Malaysia: Policy implications. *Asia and Pacific Migration Journal*, 7(2-3), 255-280.

Piore M.J. 1979. *Birds of passages: Migrants labour in industrial societies*. Cambridge: Cambridge University Press.

Piper N. 2003. Bridging gender, migration and governance: Theoretical possibilities in the Asian context. *Asian Pacific Migration Journal*, 12(1-2), 1-48.

Piper N. 2004. Rights of foreign workers and the politics of migration in South-East and East Asia. *International Migration*, 42(5).

Piper N. 2005. *Gender and migration*. Policy Analysis and Research Programme. The Global Commission on International Migration. Geneva.

Piper N. 2006. *Migrant labour in Southeast Asia: Malaysia*. Asia Research Institute, NUS. Singapore: FES Project on Migrant Labour in Southeast Asia.

Pope C., Ziebland S. and Mays N. 2000. Qualitative research in health care: Analysing qualitative data. *British Medical Journal*, 320, 114-116.

Portes A. and Sensenbrenner J. 1993. Embeddedness and immigration: Notes of the social determinants of economic action. *American Journal of Sociology*, 98(6), 1320-50.

Prothero R.M. 2001. Migration and malaria risk. *Health, Risk and Society*, 2, 19-38.

Puri S. and Ritzema T. 2004. *Migrant worker's remittances, micro-finance and the informal economy: Prospects and Issues*. Working Paper, 21. Japan: GDRC.

Pyle L.J. 2001. Sex, maids and export processing: Risks and reasons for gendered global production networks. *International Journal of Politics, Culture and Society*, 15(1), 55-76.

Quibria M.G. 1986. Migrant workers and remittances: Issues for Asian developing countries. *Asian Development Review*, 4.

Quibria M.G. 1989. International migration and real wages: Is there any Neo-classical ambiguity? *Journal of Development Economics*, 31, 177-183.

Quibria M.G. and M. Thant. 1988. International labour migration, emigrants' remittances and Asian developing countries: Economic analysis and policy issues. In *Research in Asian economic studies*, edited by M. Dulta. Greenwich, Conn: Jal Press, 287-311.

Raghuram P. 2000. Gendering skilled migratory streams: Implications for conceptualizations of migration. *Asian Pacific Migration Journal*, 9 (4), 429-457.

Rahim A.B.M. 2002. Future of labour export to Saudi Arabia. In *Future of labour export to Saudi Arabia*, edited by T. Siddiqui. Dhaka: RMMRU.

Rahman M.A.K.M. and Caples S. 1991. Devaluation, temporary migration and the labour exporting economy. *Journal of Economics and Business*, 43, 157-164.

Rahman M.M.D. 2004. Migration and poverty in Bangladesh: Ironies and paradoxes. In *International labour migration from South Asia*, edited by O.

Hisaya, Institute of Developing Economies, JETRO, Japan, ASEDP 70, 181-208.

Rahman M.M.D. 2004a. Migration networks: An analysis of Bangladeshi migration to Singapore. *Asian Profile*, 32(4), 367-390.

Rahman R. 2003. Bangladesh. Those the migrants left behind women's feature service National policy on HIV a must. *The Independent*. Dhaka, Tuesday, 14 January.

Rahn W.M., Krosnick J.A. and Breuning M. 1994. Rationalization and derivation processes in survey studies of political evaluation. *American Journal of Political Science*, 38(3), 582-600.

Ramachandran S. 2002. Operation pushback: Sangh Parivar, state, slums and surreptitious Bangladeshi in New Delhi. *Singapore Journal of Tropical Geography*, 23(3), 311-332.

Ramachandran S. 2005. Indifference, impotence and intolerance: Transnational Bangladeshis in India. *Global Migration Perspectives*, 42, Geneva, September.

Ranis G. and Fei J.C.H. 1996. A theory of economic development. *American Economic Review*, 51, 533-565.

Ravenstein E. 1885. The laws of migration. *Journal of the Statistical Society*, 46, 167-235.

Rivera B.F. 1984. International migration, non-traded goods and economic welfare in a two class economy. *Journal of Development Economics*, 16, 325-330.

Robinson K. 2000. International labour migration of Asian women. Asian Studies Association of Australia. *Asian Studies Review*, 24(2).

Roemer J.E. 1988. *Free to lose: An introduction to Marxist economic philosophy*. Cambridge: Harvard University Press.

Roy K. 1999. Low-income single fathers in an African American community and the requirement of welfare system. *Journal of Family Issues*, 20(4).

Rudnick A. 1995. Bangladeshi workers in the textile industry in Penang: People or commodities? *Global Media Newsletter*. Amsterdam, 4 October.

Rushd A. 2004. Remittances play important role in country's economy. *Daily Inqilab*. Dhaka, 28 July.

Russell S.S. 1986. Remittances from international migration: A review in perspective. *World Development*, 14(6), 677-696.

Russell S.S. 1992. Migrant remittances and development. *International Migration Quarterly Review*, 30(3/4), 267-287.

Russell S.S. 1995. *The impact of international migration on sending countries*. State Department Conference on Latin American Migration: The Foreign Policy Dilemma.

Sadler B. and Jacobs P. 1989. A key to tomorrow: On the relationship of environmental assessment and sustainable development. In *Sustainable development and environmental assessment: Perspectives on planning for a common future*, edited by P. Jacobs and B. Sadler, Montreal: Canadian Environmental Assessment Research Council.

Saith A. 1989. Macroeconomic issues in international labour migration: A review. In *To the Gulf and back: Studies on the economic impact of Asian labour migration*, edited by R. Amjad. New Delhi: ILO/ARTEP, 28-48.

Saith A. 1999. Migration processes and policies: Some Asian perspectives. *Asian Pacific Migration Journal*, 8(3), 285-311.

Salaff J.W. 1997. The gendered social organization of migration as work. *Asian Pacific Migration Journal*, 6(3-4), 295-316.

Saleh A. 2007. The bubble boys, the Green Zone and the 99%. *Forum*, 2(1). Dhaka, January.

Samren (South Asia Migration Resource Network). 2007. *Nepal: Facts and figures*. Dhaka.

Sassen S. 1984. Notes on the incorporation of Third World women into wage-labour through immigration and offshore production. *International Migration Review*, XVIII(4).

Sassen S. 1988. *The mobility of labour and capital: A study in international investment and labour flow*. Cambridge: Cambridge University Press.

Seddon D., Adhikari J.G. 2002. Foreign labour migration and the remittance economy of Nepal. *Critical Asian Studies*, 34(1), 19-40.

Shamim I. 2000. Trafficking in Women and Children: A Human Rights Concern. In *On the margin: Refugees, migrants and minorities*, edited by C.R. Abrar. Dhaka: RMMRU.

Shaptahik2000. 2005. The horror in the Sahara desert. *Shaptahik2000*, 7(45). Dhaka, 1 April.

Shari I. 2000. Economic growth and income inequality in Malaysia, 1971-95. *Journal of the Asia Pacific Economy*, 5(1/2), 112-124.

Shuval J. 2000. Diaspora migration: Definitional ambiguities and the theoretical paradigm. *International Migration*, 38(5), 41-55.

Siddiqui K. 2000. *Jagatpur 1977-97: Poverty and social change in rural Bangladesh*. Dhaka: University Press Limited.

Siddiqui T. 2001. *Short-term international labour migration of women from Bangladesh*. Dhaka: IOM/INSTRAW.

Siddiqui T. 2001. *Transcending boundaries: Labour migration of women from Bangladesh*. Dhaka: University Press Limited.

Siddiqui T. and Abrar C.R. 2003. *Migrant worker remittances and micro-finance in Bangladesh*. Social Finance Programme. Working Paper, 38. Dhaka: RMMRU and International Labour Office.

Siddiqui T. et al. 2004. *Work condition of Bangladeshi factory workers in the Middle Eastern Countries*. Solidarity Center Sri Lanka. Colombo.

Siddiqui T., Malik S. and Abrar C.R. 1999. *Labour migration from Bangladesh and the trade unions*. ILO.

Silvey R.M. 2000. Diasporic subjects: Gender and mobility in South Sulawesi. *Women's Studies International Forum*, 23(4), 501-515.

Silvey R.M. 2001. Migration under crisis; household safety nets in Indonesia's economic collapse. *Geoforum*, 32, 33-45.

Simic O. 2004. Victims of trafficking for forced prostitution: Protection mechanisms and the right to remain in the destination countries. *Global Migration Perspectives*, 2, Geneva.

Sinclair C.K. and Birks J.S. 2001. Migration and development: The changing perspective of the poor Arab countries. *Journal of International Affairs*, 33(2), 285-309.

Siwar C. and Kasim Y.M. 1997. Urban development and urban poverty in Malaysia. *International Journal of Social Economics*, 24(12), 1524-1535.

Sjaastad L.A. 1962. The costs and returns of human migration. *Journal of Political Economy*, 70(5), 80-93.

Skeldon R. 1990. Emigration and the future of Hong Kong. *Pacific Affairs*, 63(4), 500-24.

Skeldon R. 1997. *Migration and development: A global perspective*. New York: Longman.

Skeldon R. 2000. Trends in international migration in the Asian and Pacific region. *International Social Science Journal*, 52(165), 369-82.

Skeldon R. 2002. Migration and poverty. *Asia-Pacific Population Journal*, 17(4).

Skeldon R. 2003. *Migration and migration policy in Asia: A synthesis of selected cases*. University of Sussex.

Smith J.K. 1983. Quantitative versus qualitative research: An attempt to clarify the issue. *Educational Researcher*, 12, 6-13.

Sorensen N. 2004. *The development dimension of migrant remittance*. Danish Institute for Development Studies, IOM Working Paper Series, 1. Department of Migration Policy. Geneva: Research and Communications.

South China Morning Post. 2003. Illegal migrants enter HK in suitcases. *South China Morning Post*. Hong Kong, 4 February.

Southwest Economy. 2004. Immigrant assimilation: Is the U.S. still a melting pot? Federal Reserve Bank of Dallas. *Southwest Economy*, 3, May/June 2004.

Spittel M. 1998. *Testing network theory thorough an analysis of migration from Mexico to the United States*. Working Paper, 90-01. Center for Demography and Ecology, University of Wisconsin-Madison.

Stahl W.C. 2003. International labour migration in East Asia: Trends and policy issues. In *Migration in the Asia Pacific: Population, settlement and citizenship issues*, edited by R. Iredale, Hawksley and S. Castles. Cheltenham: Edward Elgar.

Stark O. and Taylor J.E. 1991. Migration incentives, migration types: The role of relative deprivation. *The Economic Journal*, 163-78, September.

Stark O., Helmenstein C. and Prskawetz A. 1997. A brain gain with a brain drain. *Economic Letters*, 55(2), 227-234.

Stouffer S. 1940. Intervening opportunities: A theory relating mobility and distance. *American Sociological Review*, 5, 845-67.

Sudman S. 1976. *Applied sampling*. New York: Academic Press.

Surtees R. 2003. Female migration and trafficking in women: The Indonesian context. *Development*, 46(3), 99-106.

Swamy G. 1981. *International migrant workers' remittances: Issues and prospects.* World Bank Staff Working Paper, 481. Washington, DC: World Bank.

Sziarto K.M. 2002. The subject(s) of migration: Cultural approaches to migration studies. In *Transnationalism, international migration, race, ethnocentrism and the State.* Graduate Student Summer Institute.

Tahmina Q.A. 2003. Weak bargaining power heightens risks for women migrants. 8th Regional Conference on Migration, Dhaka. *Migrant Forum in Asia* (MFA), Asian Migrant Centre and the Welfare Association for Repatriated Bangladeshi Employees (WARBE).

Tambiah Y. 2002. *Women and governance in South Asia: Re-imaging the State.* Colombo: ICES.

Taran P.A. 2000. Human rights of migrants: Challenges of the new decade. *International Migration*, 38(6), 7-51.

Taylor R. 2003. Hardship at home or hardship abroad: The migration system does not work. *UN Chronicle*, No. 1.

Thakur S. and Kishtwaria J. 2003. Pattern and determinants of migration: A study of Kangra district of Himachal Pradesh. *Journal of Human Ecology*, 4(4), 281-285.

Thomas R.A.L. 2005. Biometrics, international migrants and human rights. *Global Migration Perspectives*, 17, Geneva, January.

Thompson H. 1984. International migration, non-traded goods and economic welfare in the source country: A comment. *Journal of development economics*, 16, 321-324.

Thouez C. 2004. The role of civil society in the migration policy debate. *Global Migration Perspectives*, 12, Geneva.

Tientrakul C. 2003. Servants of globalization: Women, migration and domestic work. *National Women's Studies Association Journal*, 15(2), 195-199.

Tisdell C. and Gopal R. 2000. Push-and-pull migration and satisficing versus optimizing migratory behavior: A review and Nepalese evidence. *Asian Pacific Migration Journal*, 9(2), 213-229.

Todaro M.P. 1969. A model of labour migration and urban unemployment in less developed courtiers. *American Economic Review*, 59(1), 138-148.

Todaro M.P. 1985. *Rural – Urban migration: Theory and policies. Economics for a developing world*, 2nd edition. London: Longman, 209-220.

Toms S. 2004. *Emotional cost of Philippine exodus Manila.* [Online] Available at: http://news.bbc.co.uk/2/hi/asia-pacific/3651246.stm [accessed 13 September 2004].

Tyner J.A. 1996. The gendering of Philippine international labour migration. *Professional Geographer*, 48(4), 405-416.

Tyner J.A. and Donaldson D. 1999. The geography of Philippines international migration fields. *Asia Pacific Viewpoint*, 40(3), 217-234.

Ullah A.K.M.A. 2001. Migrants in Dhaka city: Life at the margins. *Asian Migrant*, XIV(3), 85-88.

Ullah A.K.M.A. 2005. Understanding the trafficking trajectories in Bangladesh: Determining essential variants. *Empowerment*, 12, 75-84.

Ullah A.K.M.A. 2008. The price of migration from Bangladesh to distant lands: Narratives of recent tragedies, *Asian Profile*, 36(6), 639-646.

Ullah A.K.M.A. 2010. Theoretical rhetoric about migration networks: A case of a journey of Bangladeshi workers to Malaysia. *International Migration*, Blackwell Wiley. Early view. 15 October 2009, 5:19am.

Ullah A.K.M.A. and Hossain M.A. 2005. Risking lives beyond borders: Reflection on an international migration scenario. In *International migration of population: Russia and contemporary world*, edited by Iontsev Vladimir. Moscow State 'Lomonosov' University. Russia: Max Press.

Ullah A.K.M.A. and Panday P.K. 2007. Remitting money to Bangladesh: What do migrants prefer? *Asian and Pacific Migration Journal*, 16(1), 121-137.

Ullah A.K.M.A., Rahman A.R. and Murshed M. 1999. *Poverty and migration: Slums of Dhaka city-the realities*. Dhaka: ARDS.

Ullah A.K.M.A. and Routray J.K. 2003. *NGOs and development: Alleviating rural poverty in Bangladesh*. Chile: Book Mark International.

UNCHR. 2000. *The sate of the world's refugees 2000: Fifty years of humanitarian action*. Oxford: Oxford University Press.

UNDP. 1996. *Human Development Report 1995*. New York: Oxford University Press.

UNDP. 1999. *Human Development Report*. New York.

UNDP. 2000. *Human Development Report*. New York.

UNDP. 2004. *Human Development Report*. New York.

UNESCAP. March-April, 2006. *Impact of cross-border migration discussed*. Population headlines, No. 311. Bangkok.

Universal currency converter. 2009. [Online] Available at: http://www.xe.com/ ucc/ [accessed 5 December 2009].

Valdez Z. May 1999. *Assimilation and ethnic entrepreneurship: Koreans and Mexicans in Los Angeles*. Working Paper, UCLA, USA.

Vete M.F. 1995. The determinants of remittance among Tongans in Auckland. *Asian Pacific Migration Journal*, 4(1).

Waddington C. 2003. *Livelihood outcomes of migration for poor people*. Working Paper, T1. Sussex: Sussex Centre for Migration Research.

Waldinger R. 1997. *Social capital or social closure: Immigrant networks in the labour market*. Working Paper, 26. Lewis Center for Regional Policy Studies, School of Public Policy and Social Research, UCLA.

WARBE. 2001. Migrant workers' rights and duties in destination countries. In *Module on labour migration process for awareness campaign through community leaders and activists*. Dhaka: RMMRU.

Wee V. and Sim A. 2003. *Transnational labour networks in female labour migration: Mediating between Southeast Asian women workers and international labour markets*. Southeast Asian Research Centre and CAPSTRANS; City University of Hong Kong. Working Paper, 49. Hong Kong.

Werner C.A. 1998. Household networks and the security of mutual indebtedness in rural Kazakstan. *Central Asian Survey*, 17(4), 597-612.

White M.J. 1990. Effects of agricultural development policies on migration in Peninsular Malaysia. *Demography*, 27(1), 97-109.

Wickramasekara P. 1996. Recent trends in temporary labour migration in Asia. In *OECD documents: Migration and the labour market in Asia-prospects to the year 2000*. France: OECD.

World Bank. 2001. *World development report 2000/2001: Attacking poverty*. Oxford: Oxford University Press.

World Bank. 2003. *International migration, remittance and poverty in developing countries*. Poverty Reduction Group MSN MC, 4-41. Washington.

World Bank. 2009. *Migration and remittance trends 2009: A better-than-expected outcome so far, but significant risks ahead*. Migration and Development Brief, 11. Washington.

Yang R. 2003. Globalization and higher education development: A critical analysis. *International Review of Education*, 49(3-4).

Young K. 2004. *Southeast Asian migrant workers in East Asian households: Globalization, social change and the double burden of market and patriarchal disciplines*. Southeast Asian Research Centre and CAPSTRANS; City University of Hong Kong. Working Paper, 58. Hong Kong.

Zafarullah H. 1998. National administration in Bangladesh: An analysis of organizational arrangements and operating methods. *Asian Journal of Public Administration*, 20(1), 79-112.

Zamir Z.B. 1998. *Socio-economic conditions of the migrant workers: A case study of Bangladeshi workers in Selangor, Malaysia*. Mimeo. Dhaka, Samren.

Zeitlyn B. June 2006. *Migration from Bangladesh to Italy and Spain*. Dhaka: RMMRU.

Zeng Z. and Xie Y. 2004. Asian-Americans' earnings disadvantage reexamined: The role of place of education. *American Journal of Sociology*, 109, 1075-1108.

Zeng Z. 2004. *The economic assimilation of Asian immigrants: A longitudinal study*. Department of Sociology. Madison: University of Wisconsin.

Zimmermann K.F. 1992. *Migration and economic development*. Munchen: Universitat Munchen.

Zlotnik H. 1992. *Asian migration to Latin America*. Proceedings of the Conference on the Peopling of the Americas, 2, 445-462. Veracruz, Mexico: IUSSP.

Zolberg A.R. 1996. The next waves: Migration theory for a changing world. In *Theories of migration*, edited by R. Cohen. London: Edward Elgar.

Index

International Labour Organization (ILO)
14, 17, 158
International Monetary Fund (IMF) 155
International Organization for Migration
(IOM) 5, 9, 14, 17, 40, 74, 97, 98,
101–2, 104, 108, 123, 124, 155,
158, 167, 169, 170, 173, 174–5,
180
International Research and Training
Institute for the Advancement of
Women (INSTRAW) 9, 17, 101–2,
124
interpersonal networks 82, 93–5, 102
investment
portfolio investment model 32
productive use of remittances 170,
171–2, 180

jail terms 133, 146
Jones, S. 16, 131

Kabeer, N. 16, 38
Kassim, A. 12–13, 15
kinship *see* family
Krieger, H. 31, 35, 137, 139

labour-export agreement, Malaysia and
Bangladesh 14, 15
language learning 43, 140–1
legal status 74–7, 145–50
see also entries beginning illegal
length of stay 73–4, 77, 117–19, 132, 133,
158
loans and repayment 104–5, 106–8, 192–3
logistic model 55

Malaysia 6–7, 12–15, 35–7
illegal workers 6–7, 13, 14, 35–8,
46–7, 87, 129–30
intermarriage 43
policy issues 218–20 *appendix*
research methods outline 46–7
routes and hardships 37–8, 85–93
marital status 69, 142, 143
marriage to local women 43
Massey, D. 29, 81, 90, 93
et al. 24, 27, 29, 33, 34, 82
Middle East 5, 9, 156

Ministry of Expatriates' Welfare and
Overseas Employment (MEWOE)
196–7
Miyan, A.M. 28, 68, 98, 100, 103–4, 108,
158, 175, 178, 179, 180
modern theory of migration 34–43
multiple regression 55, 114–16
Muslims 60
Myanmar 6–7
Myles, J. and Hou, F. 17

Nasra, S.M. 75
and Menon, I. 35, 36, 93
National Board of Revenue (NBR),
Bangladesh 197
nationality of employers 130–1
neo-classical economic macro-theory of
migration 29–34, 188–9
networks 34–8, 189, 191–2
interpersonal 82, 93–5, 102
post-migration communication and
home visits 150–3
types and scale of 82–4
see also recruiting networks
Nikolinakos, M. 26, 30
non-governmental organizations (NGOs),
as funding sources 104

occupational illnesses 149
Oishi, N. 17, 33, 40, 41

Pakistan 6–7, 13
passports 78, 86, 87, 145, 154
see also visas/permits
Philippines 6, 7, 11, 13
Piper, N. 13, 39, 75, 87, 124
police 46–7, 58, 71, 128, 140, 141
bribes 43, 129, 142, 145, 147, 153
passport/visa checks 145, 147, 154
raids 146, 147
policy issues 195–8, 217–21 *appendix*
population growth, Bangladesh 7, 8
portfolio investment model 32
poverty
and female migration 40
as push force for migration 7–8, 27
and relative deprivation 31–2, 189
prestige 109–11

For Product Safety Concerns and Information please contact our EU representative GPSR@taylorandfrancis.com Taylor & Francis Verlag GmbH, Kaufingerstraße 22, 80331 München, Germany